PSpice for Linear Circuits

BICENTENNIAL
1807
⊕WILEY
2007
BICENTENNIAL

THE WILEY BICENTENNIAL—KNOWLEDGE FOR GENERATIONS

*E*ach generation has its unique needs and aspirations. When Charles Wiley first opened his small printing shop in lower Manhattan in 1807, it was a generation of boundless potential searching for an identity. And we were there, helping to define a new American literary tradition. Over half a century later, in the midst of the Second Industrial Revolution, it was a generation focused on building the future. Once again, we were there, supplying the critical scientific, technical, and engineering knowledge that helped frame the world. Throughout the 20th Century, and into the new millennium, nations began to reach out beyond their own borders and a new international community was born. Wiley was there, expanding its operations around the world to enable a global exchange of ideas, opinions, and know-how.

For 200 years, Wiley has been an integral part of each generation's journey, enabling the flow of information and understanding necessary to meet their needs and fulfill their aspirations. Today, bold new technologies are changing the way we live and learn. Wiley will be there, providing you the must-have knowledge you need to imagine new worlds, new possibilities, and new opportunities.

Generations come and go, but you can always count on Wiley to provide you the knowledge you need, when and where you need it!

WILLIAM J. PESCE
PRESIDENT AND CHIEF EXECUTIVE OFFICER

PETER BOOTH WILEY
CHAIRMAN OF THE BOARD

PSpice for Linear Circuits

Second Edition

JAMES A. SVOBODA
Clarkson University

JOHN WILEY & SONS, INC

EXECUTIVE PUBLISHER	Don Fowley
ASSOCIATE PUBLISHER	Dan Sayre
SENIOR ACQUISITIONS EDITOR	Catherine Fields Shultz
PROJECT EDITOR	Gladys Soto
EDITORIAL ASSISTANT	Chelsee Pengal
SENIOR PRODUCTION EDITOR	Ken Santor
COVER DESIGNER	Michael St. Martine
BICENTENNIAL LOGO DESIGN	Richard J. Pacifico

This book was set in Microsoft Word® by the author and printed and bound by Courier Digital Solutions. The cover was printed by Courier Digital Solutions.

This book is printed on acid free paper. ∞

ISBN-13   978-0-471-78146-2

Printed in the United States of America

10 9 8 7 6 5

Preface

PSpice® and OrCAD® Capture are computer programs that simulate electric circuits. This manual provides an introduction to these programs and describes ways in which they can be profitably used in an introductory course on electric circuits.

This manual provides step-by-step instruction for using PSpice and OrCAD Capture to

1. analyze dc circuits, including variable dc circuits;
2. analyze ac circuits;
3. analyze circuits in the time domain to determine the complete response; and
4. analyze circuits in the frequency domain to determine the frequency response.

A formal problem solving procedure is described in Chapter 1 of this manual and used throughout the manual. Verifying the correctness of simulation results is an important step in this problem solving procedure. Consequently, every example in this manual explicitly examines the computer output to see if it is correct. A variety of techniques for checking answers are illustrated and the discipline of always checking answers is emphasized. Usually, the simulation results are correct. Occasionally they are not, and need to be corrected.

PSpice® and OrCAD® Capture are available on the Cadence OrCAD 15.7 Demo CD. A copy of this CD has been packaged with this manual. The CD is also available from the Cadence Web site using the URL www.OrCAD.com.

To install PSpice and OrCAD Capture, put the Cadence OrCAD 15.7 Demo CD into the CD player and run the setup program. (For example, from the Start Menu, select Run and then enter the filename E:\setup.exe where E is the letter assigned to the CD drive.)

After installation, Capture can be run from the Start Menu by selecting:

Start / All Programs / OrCAD 15.7 Demo / OrCAD Capture CIS Demo.

Contents

CHAPTER 1

Getting Started with PSpice

SPICE is a computer program used for numerical analysis of electric circuits. SPICE is an acronym for Simulation Program with Integrated Circuit Emphasis. Developed in the early 1970's at the University of California at Berkley, SPICE is generally regarded to be the most widely used circuit simulation program [1]. PSpice is a version of SPICE for personal computers developed by MicroSim Corporation in 1984 [2]. SPICE was a text based program. The user was required to describe the circuit using only text, and the simulation results were displayed as text. MicroSim provided a graphical postprocessor, Probe, to plot the results of SPICE simulations. Later, MicroSim also provided a graphical interface called Schematics that allowed users to describe circuits graphically. The name of the simulation program was changed from PSpice to PSpice A/D when it became possible to simulate circuits that contained both analog and digital devices. MicroSim was acquired by OrCAD, which was in turn acquired by Cadence. OrCAD improved Schematics and renamed it Capture. "Using PSpice" loosely refers to using OrCAD Capture, PSpice A/D and Probe to numerically analyze an electric circuit.

1.1 Using PSpice

In this manual, a six step procedure will be used to organize circuit analysis using PSpice. This procedure is illustrated in Figure 1.1 and is stated as follows.

Step 1 Formulate a circuit analysis problem.

Step 2 Describe the circuit using OrCAD Capture. This description requires three activities.
1. Place the circuit elements in the OrCAD Capture workspace.
2. Adjust the values of the circuit element parameters, e.g. the resistances of the resistors.
3. Wire the circuit to connect the circuit elements.

Step 3 Simulate the circuit using PSpice.

Step 4 Display the results of the simulation, for example, using Probe.

Step 5 Verify that the simulation results are correct.

Step 6 Report the answer to the circuit analysis problem.

The second, third and fourth steps of this procedure use the software. In the first and fifth steps, the user identifies the problem that is to be solved and verifies that it has indeed been solved. It would be hard to overemphasize the importance of steps 1 and 5. If the simulation is not correct, the error must be identified and eliminated so that a correct simulation can be performed. Once a correct simulation has been performed, the answer can be reported.

Example 1.1 This example illustrates the six step circuit analysis procedure and introduces the programs OrCAD Capture and PSpice A/D.

Step 1 Formulate a circuit analysis problem.

Analyze the circuit shown in Figure 1.2 to determine the value of $v(t)$, the voltage across the 6 Ω resistor.

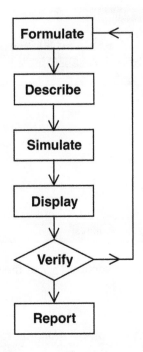

Figure 1.1 A six step procedure for using PSpice.

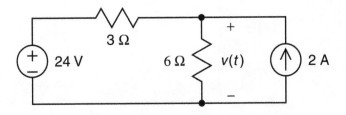

Figure 1.2 The circuit for Example 1.1.

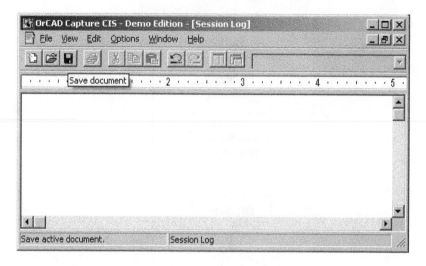

Figure 1.3 The opening screen of version 15.7 of the OrCAD Capture CIS Demo.

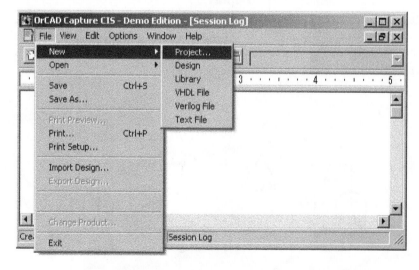

Figure 1.4 Opening a new project in OrCAD Capture.

Step 2 Describe the circuit using OrCAD Capture.

Begin by starting the OrCAD Capture program. Figure 1.3 shows the opening screen of OrCAD Capture. (If necessary, maximize the "Session Log" window.) The top line of the screen shows the title of the program, "OrCAD Capture CIS – Demo Edition." A menu bar providing menus called File, View, Edit, Options, Window, and Help is located under the title line. A row of buttons is located under the menu bar and a ruler is located below the row of buttons. A workspace is located beneath the ruler. The circuit to be simulated is described by drawing it in this workspace. A line containing two message fields is located under the workspace. The left message field is of particular interest. This message field provides information about the Capture screen. For example, move the mouse cursor to one of the buttons. The left message field describes the function of the button. "Save active document" is the function of the third button from the left.

Select **File / New / Project** from the Capture menus as shown in Figure 1.4. The "New Project" dialog box, shown in Figure 1.5, will pop up. Select "Analog or Mixed A/D" as shown. The "New Project" dialog box requires a project name and a location. The location is the name of the directory or folder where Capture should store the project file. The name will be the file name of the project file. OrCAD Capture uses opj as a suffix for project files, so choosing Name to be "first example" and Location to be "c:\SPICE Manual\chap1" causes OrCAD to store a file "first example.opj" in the folder "c:\SPICE Manual\chap1." Notice that long file names are supported, making it easier to give descriptive names to projects.

Select "OK" in the "New Project" dialog box to close the "New Project" dialog box and pop up the "Create PSpice Project" dialog box shown in Figure 1.6. Select "Create a blank project" and then "OK" to return to the OrCAD Capture screen shown in Figure 1.7. The Capture screen has changed. "Place," "Macro," "PSpice" and "Accessories" have been added to the menu bar, there are more buttons, and there is a grid on the workspace.

Drawing a circuit requires three activities.

1. Place the circuit elements in the OrCAD Capture workspace.

2. Adjust the values of the circuit element parameters, e.g., the resistances of the resistors.

3. Wire the circuit to connect the circuit elements.

Select and place the parts that comprise the circuit.

Select **Part / Place** from the Capture menus to pop up the "Place Part" dialog box shown in Figure 1.8. Click on the "Add Library…" button to pop up the dialog box shown in Figure 1.9. OrCAD Capture provides several libraries containing parts for circuits. File names of parts libraries use the suffix "olb." Figure 1.9 shows the libraries provided in the PSpice folder. Select the libraries "analog.olb" and "source.olb." Click on the "Open" button to make these libraries available and to return to the "Place Part" dialog box as shown in Figure 1.10.

To obtain a resistor, select "ANALOG" from the list of Libraries and R from the list of Parts. Select "OK" to close the "Place Part" dialog box and return to the Capture screen.

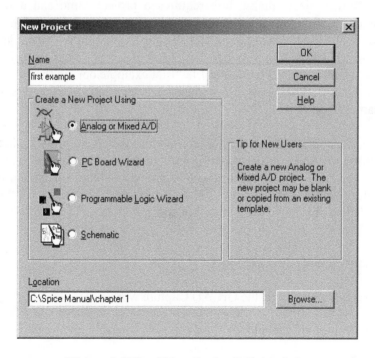

Figure 1.5 The "New Project" dialog box.

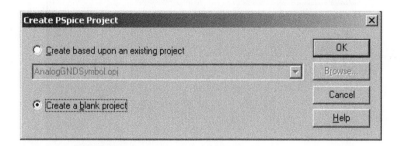

Figure 1.6 Create PSpice Project dialog box.

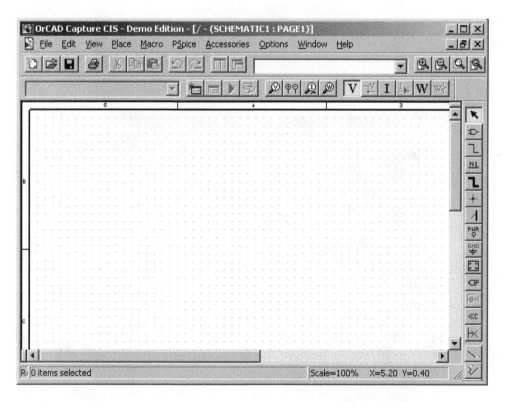

Figure 1.7 A new project displayed in the Capture window.

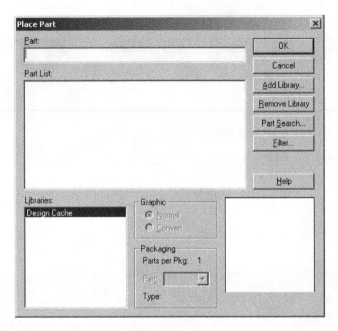

Figure 1.8 The Place Part dialog box before adding parts libraries.

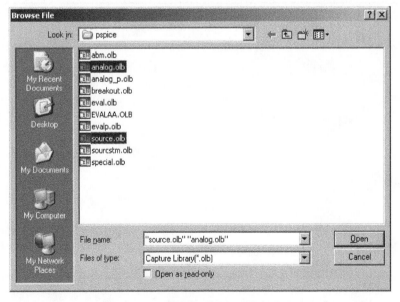

Figure 1.9 Adding parts libraries.

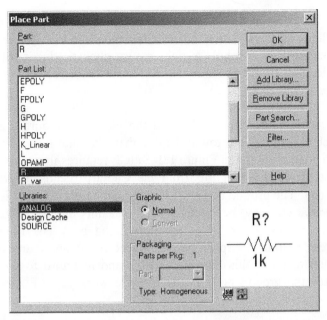

Figure 1.10 The Place Part dialog box after adding parts libraries.

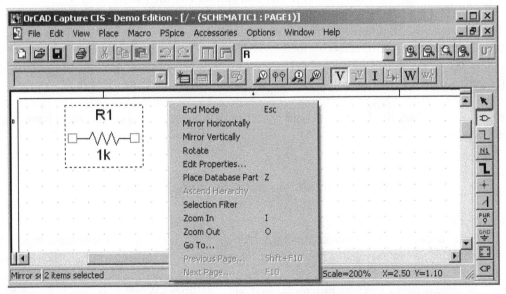

Figure 1.11 A right mouse click while placing parts pops up this menu.

Upon returning to the Capture screen, the cursor will be dragging the symbol for a resistor. Place the resistor as desired using a left mouse click. The cursor will now be dragging a second resistor symbol. A right mouse click produces the menu shown in Figure 1.11. Selections from this menu will flip or rotate the resistor. Select "End Mode" to stop placing resistors.

Table 1.1 lists the parts needed to simulate the circuit in Figure 1.2, as well as the libraries in which these parts are found. Notice that a ground node is included in Table 1.1, even though none is shown in Figure 1.2. SPICE requires that every circuit include a ground node. **Select Part / Ground** from the Capture menus to pop up the "Place Ground" dialog box. The ground node is a PSpice part called "0" that is contained in the "SOURCE" library. (It may be necessary to add this library. Click on the "Add Library…" button to pop up a "Browse File" dialog box. The library file is called source.olb and resides in the "PSpice" folder. Select the library source.olb, then click on the "Open" button to make this library available and to return to the "Place Ground" dialog box.) Place the ground node in the Capture workspace. Figure 1.12 shows the Capture screen after the parts have been placed.

Adjust the values of the parameters of the parts.

The parameters of the resistors each have their default values, 1k. Left click on the "1k" of the vertical resistor to select it, then right click anywhere in the Capture workspace to obtain the menu shown in Figure 1.13. Choose "Edit Properties" to pop up the Display properties dialog box shown in Figure 1.14. Change the value from 1k to 6.

Figure 1.15 shows the Capture screen after the parameter values of the parts have been adjusted.

Wire the circuit to connect the circuit elements.

Select **Parts / Wire** to wire the parts together. Consider Figure 1.15. Notice that the terminals of each part are marked using small squares. To wire two terminals together, left-press the mouse on one terminal, drag the mouse to the other terminal, then release the mouse. The path of the wire will generally follow the path of the mouse, but wires will be drawn using straight horizontal and vertical lines. Wires can also connect part terminals to wires, or connect wires to wires.

To stop wiring, right-click the mouse, then select "End Mode" from the menu that appears. Figure 1.16 shows the circuit after it has been wired.

Table 1.1 PSpice parts and the libraries in which they are found

Symbol	Description	PSpice Name	Library
R? — 1k	resistor	R	ANALOG
0Vdc V?	DC voltage source	VDC	SOURCE
0Adc I?	DC current source	IDC	SOURCE
0	ground (reference node)	0	SOURCE

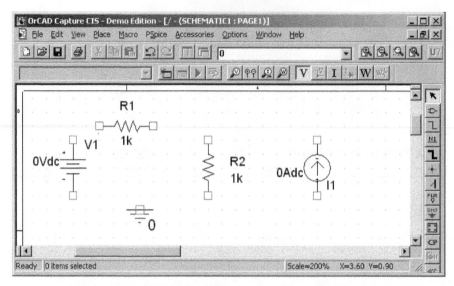

Figure 1.12 OrCAD Capture screen after placing the parts.

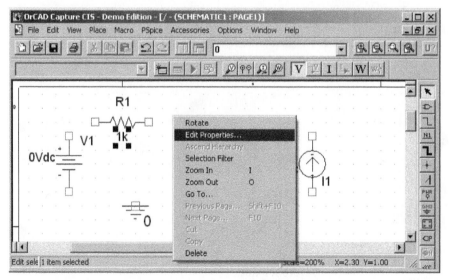

Figure 1.13 The value "1k" is shown highlighted. Right-clicking anywhere in the Capture workspace pops up this menu.

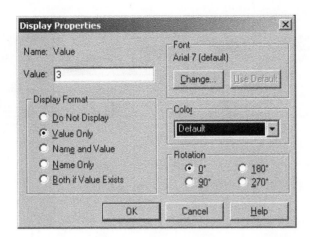

Figure 1.14 The Display Properties dialog box.

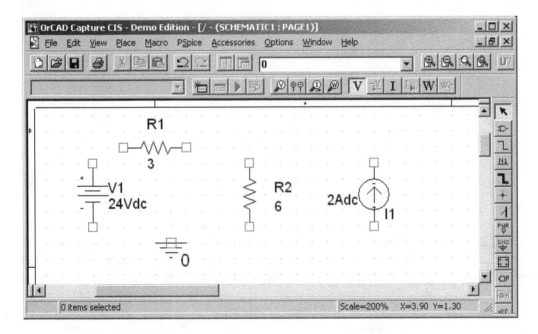

Figure 1.15 Capture screen after adjusting the values of the circuit parameters.

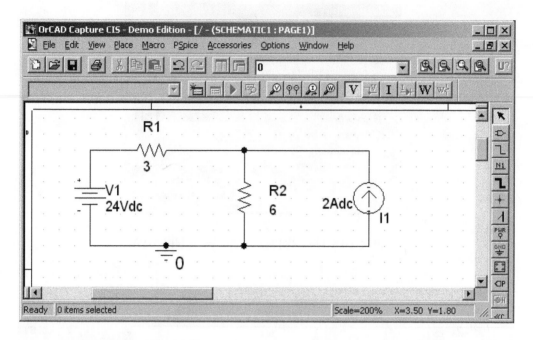

Figure 1.16 The circuit of Figure 1.2 as described in Capture.

Step 3 Simulate the circuit using PSpice.

Select **PSpice / New Simulation Profile** from the OrCAD Capture menus to pop up the New Simulation dialog box shown in Figure 1.17. Provide a Name such as "dCkt" then select Create. The "Simulation Settings" dialog box shown in Figure 1.18 will pop up. Select "Bias Point" from the Analysis type list and check "General Settings" under options. Select "OK" to close the "Simulation Settings" dialog box. Select **PSpice / Run** as shown in Figure 1.19 to run the simulation and pop up the Schematics window.

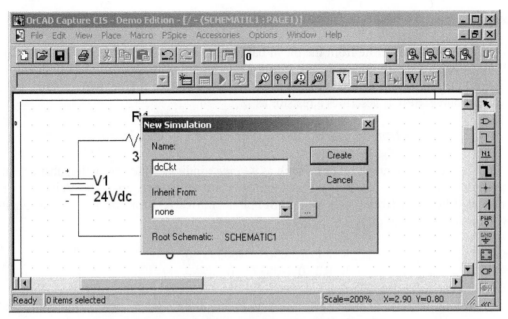

Figure 1.17 New Simulation dialog box.

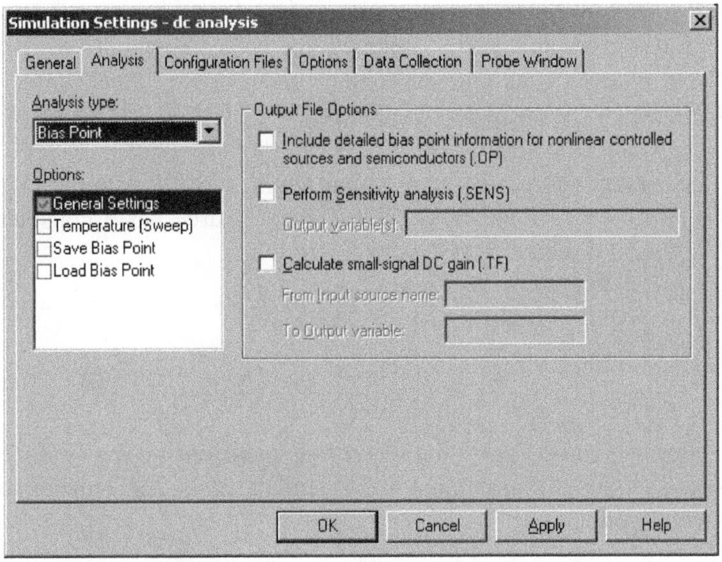

Figure 1.18 The "Simulation Settings" dialog box.

Step 4 Display the results of the simulation.

Select **View / Output File** from the menu bar at the top of the Schematics window. The PSpice output file will appear in the scrolled window. Scroll down to the list shown in Figure 1.20. This is a list of the devices in the circuit, a current source, two resistors and a voltage source. SPICE describes a resistor as

$$\textit{<Name>}\quad \textit{<Node1>}\quad \textit{<Node2>}\quad \textit{<Value>}$$

(An italic font inside angle brackets such as *<Name>* indicates the user will substitute an appropriate name for "Name.") This resistor has a resistance equal to Value and is connected between Node1 and Node2. SPICE identifies resistors by names that begin with the letter R. The lines

$$\text{R\_R1} \quad \text{N00365} \quad \text{N00372} \quad 3$$
$$\text{R\_R2} \quad 0 \quad \text{N00372} \quad 6$$

describe a 3Ω resistor connected between node 365 and node 372 and a 6Ω connected between nodes 0 and 372. (OrCAD Capture has added R_ to the beginning of the name R1 and R2 used for the resistors in Figure 1.16, insuring that the names of the resistors begin with R.)

Similarly, SPICE describes a dc current source as

$$\textit{<Name>}\quad \textit{<Node1>}\quad \textit{<Node2>}\quad [DC]\quad \textit{<Value>}$$

SPICE identifies current sources by names that begin with the letter I. (Square brackets such as [DC] indicate an optional field.) The current source has a constant current with size equal to Value and directed from Node1 to Node2.

The third line

$$\text{I\_I1} \quad 0 \quad \text{N00372} \quad \text{DC} \quad \text{2Adc}$$

describes the current source. This line describes a current source with Name = I_I1, Node1 = 0, Node2 = N00372 and value = 2 A, indicating a 2 A current source directed from node 0 toward node 372.

17

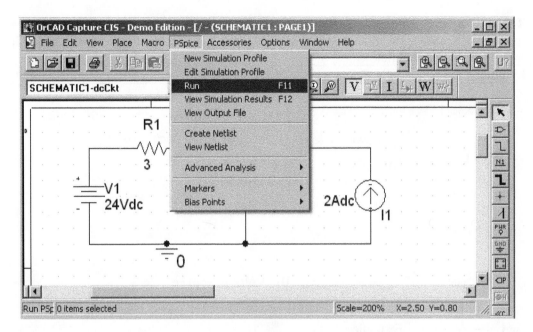

Figure 1.19 Run the simulation.

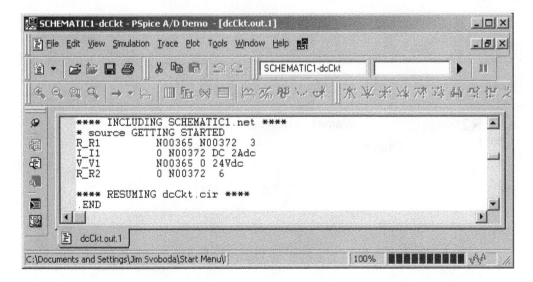

Figure 1.20 The PSpice output file.

Table 1.2 SPICE Syntax. An italic font inside angle brackets such as *<Name>* indicates the user will substitute an appropriate name for "Name." Square brackets such as [DC] indicate the a optional field

Symbol	SPICE Syntax

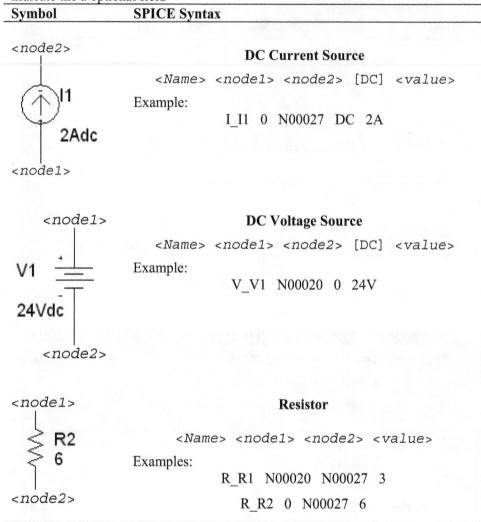

DC Current Source

<Name> <node1> <node2> [DC] *<value>*

Example:

    I_I1   0   N00027   DC   2A

DC Voltage Source

<Name> <node1> <node2> [DC] *<value>*

Example:

    V_V1   N00020   0   24V

Resistor

<Name> <node1> <node2> <value>

Examples:

    R_R1   N00020   N00027   3

    R_R2   0   N00027   6

SPICE describes a dc voltage source as

<Name> <Node1> <Node2> [DC] <Value>

This voltage source has a constant voltage with size equal to Value and with polarity marked with a + at Node1 and a − at Node2. SPICE identifies voltage sources by names that begin with the letter V. The line

V_V1 N00365 0 24Vdc

describes a voltage source with nodes 365 (+) and 0 (−) and having a constant voltage value of 24 V.

Figure 1.21 shows the circuit after the nodes have been numbered using the node numbers generated by PSpice. Table 1.2 summarizes the SPICE syntax for resistors and dc sources. Figure 1.22 shows the Capture workspace after running the PSpice simulation. The node voltages of the circuit have been labeled using reverse video.

Scroll further down the PSpice output as shown in Figure 1.23. The PSpice output file indicates that

 a. The node voltage at node 365 is 24 V.
 b. The node voltage at node 372 is 20 V.
 c. The current in the voltage source is −1.333 A.
 d. The total power dissipation is 32 W.

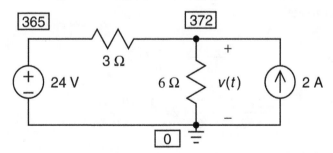

Figure 1.21 The circuit from Figure 1.2 after numbering the nodes using the node numbers generated by PSpice.

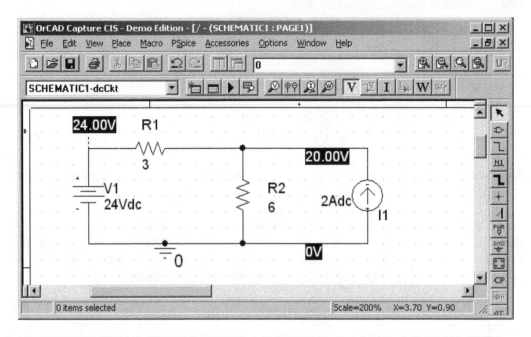

Figure 1.22 OrCAD Capture labels node voltages after performing a PSpice simulation.

Step 5 Verify that the simulation results are correct.

To check the results of this simulation, apply Kirchhoff's Current Law (KCL) at node 372 in Figure 1.21 to get

$$\frac{v_{365} - v_{372}}{3} - \frac{v_{372}}{6} + 2 = 0$$

or

$$\frac{24 - 20}{3} - \frac{20}{6} + 2 = \frac{4}{3} - \frac{10}{3} + \frac{6}{3} = 0$$

This equation is satisfied so the simulation is likely to be correct.

SPICE uses the passive convention for all devices, including sources, so the current in the voltage source is the current directed from the + node, node 365 in this case, toward the – node, node 0 in this case. Apply KCL at node 365 to calculate the voltage source current as

$$\frac{v_{372} - v_{365}}{3} = \frac{20 - 24}{3} = -1.33 \ A$$

This is the value given in the PSpice output, making it very likely that the simulation is correct.

Finally, what is meant by the total power dissipation? The power supplied by the dc voltage source is

$$vi = 24 \left(\frac{v_{365} - v_{372}}{R_1} \right) = 24 \left(\frac{24 - 20}{3} \right) = 32 \ W$$

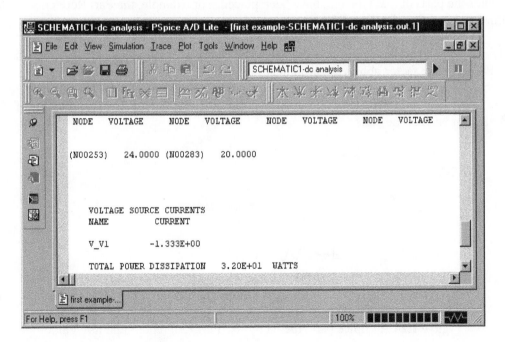

Figure 1.23 The PSpice output file.

This power, the power supplied to the rest of the circuit by all the **dc voltage** sources, is the power reported by SPICE as the total power dissipation. (The power supplied to the rest of the circuit by all the dc voltage sources is important when designing amplifiers using transistors.)

Step 6 Report the result.

The value of $v(t)$, the voltage across the 6 Ω resistor in Figure 1.2, is 20 V.

1.2 PSpice Notation

Figure 1.24 show a PSpice part, in this case, a resistor. The part is labeled twice, by a "Part Reference" and also by a "Value." The part reference consists of a letter and a number. The letter corresponds to the type of the part as shown in Table 1.3. The number counts the parts of that type that have been placed. For example, the Part Reference of the third resistor placed will be R3.

The value is a real number or an expression. In Figure 1.24 the value is "1k" where "k" is a PSpice scale factor abbreviation indicating 1000. That is, the value is 1*1000 = 1000. Table 1.4 lists the PSpice scale factor abbreviations. The "Value" of a resistor represents the resistance of the resistor. The third column of Table 1.3 gives interpretation of the Value for each of the parts.

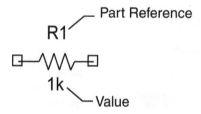

Figure 1.24 A PSpice part labeled by its "Part Reference" and its "Value"

Braces are used to indicate that the Value is an expression rather than a number. For example, if the Value of a resistor is {1000*2*3.14159} the resistance is obtained by evaluating the expression 1000*2*3.14159.

Both the Part Reference and the Value are properties of PSpice parts and are specified and changed using the Property Editor. For example to change the resistance from "1k" to "2k," left-click on the a part to select it, then right click anywhere in the Capture workspace to pop up a menu as shown in Figure 1.11. Select "Edit Properties..." from that menu to open the "Property Editor" as shown in Figure 1.25. Scroll through the list of properties to find the property of interest, "Value" in this case. Edit the value of the property (yes, the value of "Value") then left-click "Apply" to make the change. Close the property editor.

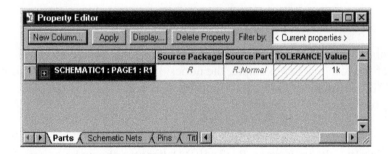

Figure 1.25 The Property Editor.

Table 1.3 PSpice parts and their labels

Part	Part Reference	Value
Current Source	Ii	source current
Voltage	Vi	source voltage
Capacitor	Ci	capacitance
Inductor	Li	inductance
Resistor	Ri	resistance
VCVS	Ei	gain
CCCS	Fi	gain
VCCS	Gi	gain
CCVS	Hi	gain
Op Amp	Ui	none

Table 1.4 PSpice scale factor abbreviations

Letter Suffix	Multiplying Factor	Name of Suffix
T	1E12	tera
G	1E9	giga
MEG	1E6	mega
K	1E3	kilo
M	1E−3	milli
MIL	25.4E−6	mil
U	1E−6	micro
N	1E−9	nano
P	1E−12	pico
F	1E−15	femto

PSpice numbers the nodes of a resistor by appending the numbers :1 and :2 to the Part Reference of the resistor. For example, the nodes of resistor R1 are R1:1 and R1:2. Figure 1.25 shows the positions of nodes R1:1 and R1:2 as resistor R1 is (*a*) placed, and rotated (*b*) once (*c*) twice and (*d*) three times.

I(R1) denotes the current in resistor R1, directed from node R1:1 to node R1:2. V(R1:2) denotes the node voltage at node R1:2. V(R1) denotes the voltage across resistor R1, V(R1) = V(R1:1)−V(R1:2). Notice that V(R1) and I(R1) adhere to the passive convention.

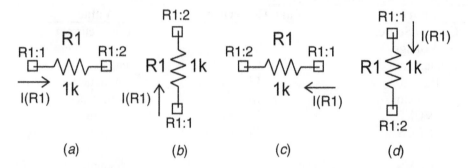

Figure 1.25 The nodes of a part.

In Example 1.1 we saw that PSpice labels nodes with labels like N00365 and N00372. These labels are not very convenient. Nodes can be given more convenient names using the using a PSpice part called an "off-page connector." Consider a particular off-page connector called an "OFFPAGELEFT-R" part and found in the part library named CAPSYM. To label a node, select **Place / Off-Page Connector…** from the OrCAD Capture menus to pop up the "Place Off-Page Connector" dialog box as shown in Figure 1.26. Select the library CAPSYM from the list of libraries. (If CAPSYM does not appear in the list of libraries, left-click the "Add Libraries" button to pop up a "Browse File" dialog box. Select capsym.olb from the list of libraries in the "Libraries" folder then left-click the "Open" button to return to "Place Off-Page Connector" dialog box.) Left-click the "OK" button to place the part in the OrCAD Capture workspace.

The new connector will be labeled as "OFFPAGELEFT-R." Use the property editor to change this name to something descriptive such as "output." (Left-click on the "OFFPAGELEFT-R" label to select it, then right-click anywhere in the OrCAD workspace to obtain the menu shown in Figure 1.13. Choose "Edit Properties" to pop up the "Display Properties" dialog box shown in Figure 1.14. Change the value from OFFPAGELEFT-R to "output.") Wire the connector to the appropriate node of the circuit to name that node "output."

Figure 1.27 shows the Capture screen after the node has been named. The output node has several names: output, R1:2, R2:2 and I1:2. Consequently, the node voltage at the output node has several names: V(output), V(R1:2), V(R2:2) and V(I1:2). The node voltage at the output node is equal to the voltage across the current source. Recall that V(I1) and I(I1) = 2 Adc adhere to the passive convention. Therefore V(I1) = V(I1:1)–V(I1:2) = 0 – V(output).

1.3 Summary

This chapter presented and illustrated a six step procedure for simulating electric circuits and introduced the programs OrCAD Capture and PSpice A/D.

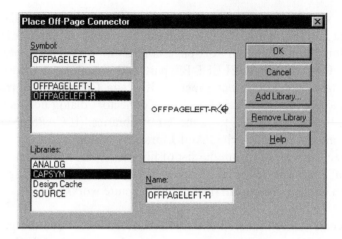

Figure 1.26 The Place Off-Page Connector dialog box.

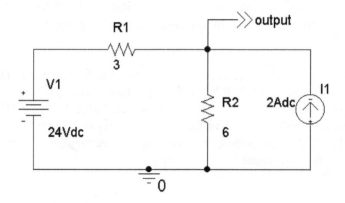

Figure 1.27 Naming a node using an off-page connecter.

1.4 References

[1] Perry, T.S. "Donald Pederson: The Father of SPICE" in the *IEEE Spectrum*, June 1998.

[2] Tuinenga, P.W. *SPICE: A Guide to Circuit Simulation & Analysis Using PSpice*, 2nd Edition, Prentice-Hall.

CHAPTER 2

Analysis of DC Circuits

A dc circuit is a circuit in which the voltages of all independent voltage sources and the currents of all independent current sources have constant values. All of the currents and voltages of a dc circuit, including mesh currents and node voltages, have constant values. PSpice can analyze a dc circuit to determine the values of the node voltages and also the values of the currents in voltage sources. PSpice uses the name "Bias Point" to describe this type of analysis. (The name "Bias Point" refers to the role of dc analysis in the analysis of a transistor amplifier.)

In this chapter we consider four examples. The first example illustrates analysis of circuits containing dependent sources. The second illustrates the use of PSpice to check the node or mesh equations of a circuit to verify that these equations are correct. The third uses PSpice to compare two dc circuits. The fourth example uses PSpice to find the Thevenin equivalent circuit of a given circuit. As always, these examples follow the six step procedure for analyzing circuits using PSpice that was introduced in Chapter 1.

2.1 DC Circuits Containing Dependent Sources

OrCAD Capture and PSpice can be used to analyze circuits that contain dependent sources. Table 2.1 shows the symbols used by OrCAD Capture to represent dependent sources.

Example 2.1 This example illustrates analysis of a circuit that contains a dependent source. Particular attention is paid to preparing the circuit for analysis using PSpice.

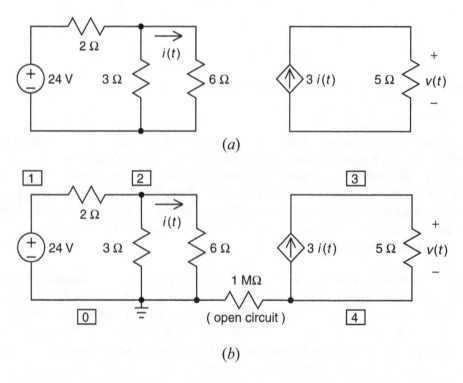

Figure 2.1 (*a*) A circuit containing a dependent source and (*b*) an equivalent circuit ready to be analyzed using PSpice.

Table 2.1 PSpice dependent sources

Symbol	Description	PSpice Name	Library
E? + − E	VCVS (Voltage Controlled Voltage Source)	E	ANALOG
F? F	CCCS (Current Controlled Current Source)	F	ANALOG
G? + − G	VCCS (Voltage Controlled Current Source)	G	ANALOG
H? H	CCVS (Current Controlled Voltage Source)	H	ANALOG

Step 1 Formulate a circuit analysis problem.

Find the voltage $v(t)$ across the 5Ω resistor in the circuit shown in Figure 2.1a.

There are two reasons why the circuit as shown in Figure 2.1a is not ready to be analyzed using PSpice. First, PSpice requires that every circuit has a reference node. Second, PSpice can only analyze connected circuits, that is, circuits that consist of a single part. The first problem can be solved by selecting the bottom left node as the reference node.

The second problem can be solved by adding an open circuit between the bottom nodes of the two parts of the circuit. Adding an open circuit will not alter the circuit shown in Figure 2.1a. In particular, adding an open circuit will not change the value of $v(t)$. Unfortunately, an open circuit is not one of the parts available in PSpice. Fortunately, a resistor with infinite resistance acts like an open circuit. An open circuit can be approximated using a resistor having a very large resistance. How large is "very large?" "Very large" means much larger than the other resistors in the circuit. A resistance of 1 MΩ is much larger than the largest resistance in the circuit shown in Figure 2.1a.

Figure 2.1b shows the circuit that will be analyzed using PSpice. That analysis will produce the values of the node voltages of the circuit. The nodes of the circuit have been numbered to make it easier to talk about the node voltages. When we verify that the simulation results are correct, we will also verify that the node voltages of the circuit shown in Figure 2.1a are the same as the node voltages of the circuit shown in Figure 2.1b.

Step 2 Describe the circuit using OrCAD Capture.

Start OrCAD Capture and create a new project as described in Chapter 1.

The circuit in Figure 2.1b contains a Current Controlled Current Source (CCCS) in addition to resistors and a voltage source. Place the voltage source and the resistors in the OrCAD Capture workspace and adjust their values as described in Chapter 1. Table 2.1 indicates that a CCCS is a part named "F" that is contained in the "ANALOG" part library. Select **Part / Place** from the Capture menus to pop up the "Place Part" dialog box. (If necessary, left-click on the "Add Libraries..." button in the "Place Part" dialog box and add the libraries "source.olb" and "analog.olb" as described in Chapter 1.) Select

ANALOG from the list of libraries and select F from the Parts List as shown in Figure 2.2.

Figure 2.3 shows the Capture screen after the CCCS has been placed. (Notice that the value of the resistance of R5 is "1Meg" rather than "1M." Table 1.4 shows a list of scale factor abbreviations available in Capture and PSpice. PSpice will interpret 1M as 1 milliohm rather than 1 megaohm.)

To adjust the gain of the CCCS, left-click on the CCCS in the Capture workspace to select the CCCS then right-click in the Capture workspace to pop up a menu as shown in Figure 2.4. Select "Edit Properties" from this menu to bring up the property page for the CCCS, shown in Figure 2.5. Use the horizontal scrollbar near the bottom of the screen to scroll through the properties to find the property named GAIN. Edit the property GAIN to set the gain of the CCCS to the value 3 as shown in Figure 2.5. Be certain to click the button "Apply" to actually set the gain of the CCCS to the value of GAIN in the property sheet.

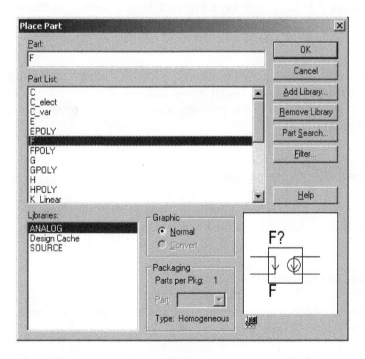

Figure 2.2 Selecting a CCCS from the ANALOG Parts Library.

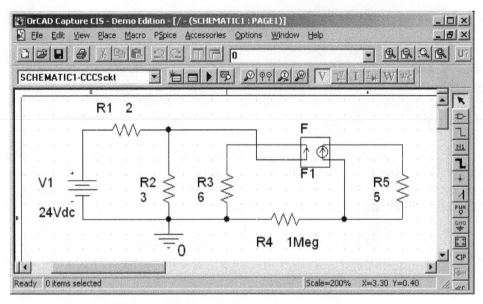

Figure 2.3 The circuit from Figure 2.1*b* as described in Capture.

Find the familiar "Minimize," "Maximize," "Close" buttons in the upper right-hand corner of the Capture window. Under these buttons is a second set of "Minimize," "Maximize," "Close" buttons for the property sheet. Close the property sheet to return to the Capture screen.

Step 3 Simulate the circuit using PSpice.

Select **PSpice / New Simulation Profile** to specify the "Analysis type" to be "Bias Point" as described in Chapter 1. Select **PSpice / Run** to run the PSpice simulation.

Step 4 Display the results of the simulation.

After the circuit has been simulated using PSpice, the Capture screen, shown in Figure 2.6, displays the node voltages in reverse video.

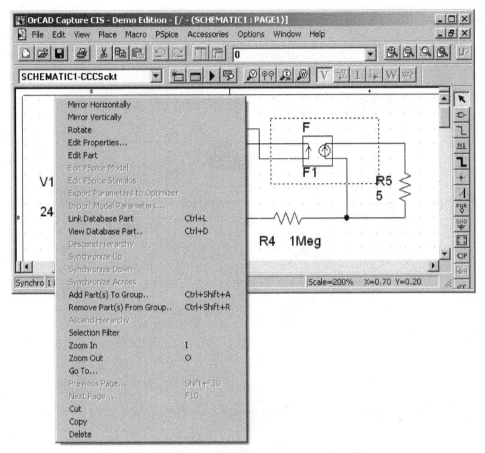

Figure 2.4 Right-click menu for the CCCS.

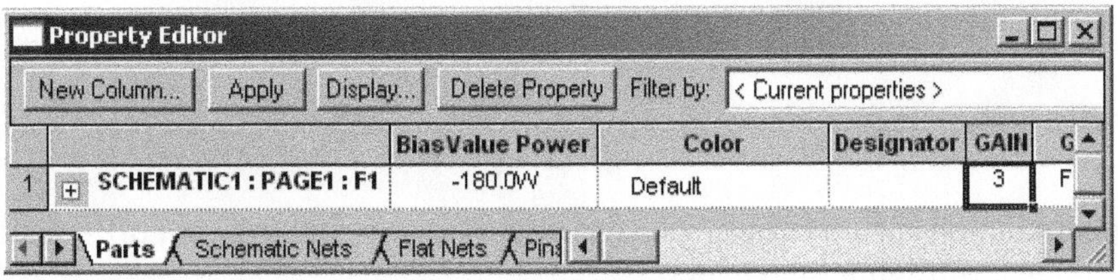

Figure 2.5 Setting the gain of the CCCS.

Step 5 Verify that the simulation results are correct.

Let v_i denote the node voltage at node i. Compare Figures 2.1b and 2.6 to see that the values of the node voltage are

$$v_1 = 24.00 \text{ V}, \quad v_2 = 12.00 \text{ V}, \quad v_3 = 30.00 \text{ V} \quad \text{and} \quad v_4 = 0 \text{ V}$$

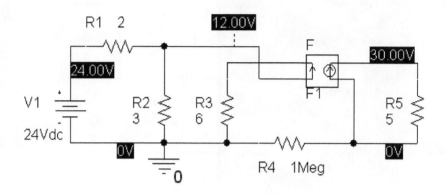

Figure 2.6 The node voltages.

Apply Kirchhoff's Current Law (KCL) at node 2 of Figure 2.1b to get

$$\frac{v_1 - v_2}{2} = \frac{v_2}{3} + \frac{v_2}{6}$$

Substituting the values of the node voltages gives

$$\frac{24 - 12}{2} = \frac{12}{3} + \frac{12}{6}$$

$$6 = 4 + 2$$

Thus the simulation results agree with the circuit. Next apply Kirchhoff's Current Law (KCL) at node 3 of Figure 2.1b to get

$$3i(t) = \frac{v_3 - v_4}{5} \quad \Rightarrow \quad 3\left(\frac{v_2}{6}\right) = \frac{v_3 - v_4}{5}$$

Substituting the values of the node voltages gives

$$3\left(\frac{12}{6}\right) = \frac{30-0}{5}$$

Again, the simulation results agree with the circuit. The simulation results are correct.

To verify that the circuits in Figures 2.1a and 2.1b are equivalent, apply KCL to node 4 of the circuit in Figure 2.1b to get

$$3\,i(t) = \frac{v_3 - v_4}{5} + \frac{0 - v_4}{1 \times 10^6}$$

The second term on the right side of this equation is the current in the 1 MΩ resistor. That current is zero. Hence the 1 MΩ resistor acts like an open circuit and the two circuits are equivalent.

Step 6 Report the result.

The value of $v(t)$, the voltage across the 5 Ω resistor in Figure 2.1a is 30 V.

2.2 Mesh and Node Equations

In this example, PSpice is used to check node and mesh equations of a circuit.

Example 2.2 Consider the circuit shown in Figure 2.7. A set of mesh currents has been labeled and the nodes of this circuit have been numbered. The circuit can be represented by the following node equations.

$$23\,v_2 - 12\,v_3 = 36$$
$$-55\,v_2 + 21\,v_3 - 6\,v_4 = 0$$
$$-6\,v_3 + 26\,v_4 = 180$$

where v_i denotes the node voltage at node i. Similarly, the circuit in Figure 2.7 can be represented by the following mesh equations.

$$17\,i_1 - 8\,i_2 - 2\,i_3 - 5\,i_4 = 0$$
$$-8\,i_1 + 11\,i_2 - 3\,i_3 = 12$$
$$-7\,i_1 - 3\,i_2 + 5\,i_3 + 11\,i_4 = 0$$
$$-4\,i_2 + 3\,i_3 + i_4 = 0$$

The objective of this example is to use PSpice to determine if these equations are correct.

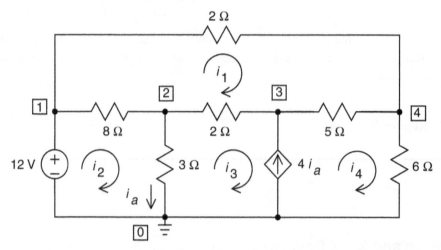

Figure 2.7 The circuit for the mesh and node equation example.

Step 1 Formulate a circuit analysis problem.

We have seen that PSpice will calculate the node voltages of a circuit such as the one shown in Figure 2.7. The node equations can be checked by determining the values of the node voltages using PSpice and substituting those values into the node equations. PSpice does not calculate mesh currents, but it does calculate the currents in voltage sources. The mesh current i_2 is the current in the 12 V voltage source. PSpice uses the passive convention for all elements, including voltage sources. The voltage source current that PSpice will report is the current directed from + to −. In this case, PSpice will report the value of $-i_2$ rather than i_2.

Similarly, zero volt voltage sources can be added to the circuit to measure the other mesh currents. A zero volt voltage source is equivalent to a short circuit. In Figure 2.8, short circuits (0 V voltage sources) have been added to the circuit to measure mesh currents i_1, i_3 and i_4.

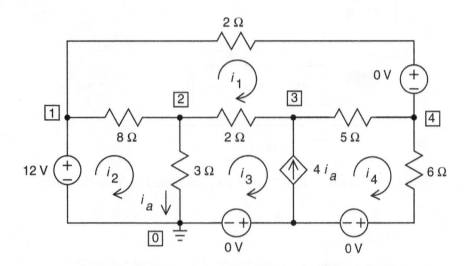

Figure 2.8 The circuit from Figure 2.7 after adding short circuits (0 V voltage sources) to measure the mesh currents.

Step 2 Describe the circuit using OrCAD Capture.

Start OrCAD Capture. Place the parts. Adjust the parameter values. Wire the circuit. Figure 2.9 shows the circuit as it was described using OrCAD Capture.

Step 3 Simulate the circuit using PSpice.

Select **PSpice / New Simulation Profile** to specify the "Analysis type" to be "Bias Point" as described in Chapter 1. Select **PSpice / Run** to run the PSpice simulation.

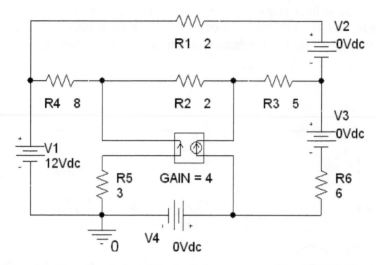

Figure 2.9 The circuit from Figure 2.8 represented in OrCAD Capture.

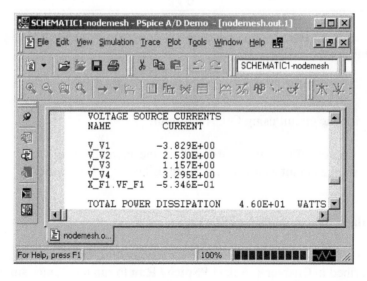

Figure 2.10 The PSpice output file shows the mesh currents.

Step 4 Display the results of the simulation.

Running the PSpice simulation will open the Schematics window. **Select View / Output File** from the menu bar at the top of the Schematics window. The PSpice output file will appear in the scrolled window. Scroll down to the voltage source currents as shown in Figure 2.10. Comparing Figures 2.8 and 2.9 shows that the current in V_V1 is $-i_2$, the current in V_V2 is i_1, the current in V_V3 is i_4 and the current in V_V4 is i_3. Thus the mesh currents are

$$i_1 = 2.530 \text{ A}, \quad i_2 = 3.829 \text{ A}, \quad i_3 = 3.295 \text{ A} \quad \text{and} \quad i_4 = 1.157 \text{ A}$$

After the circuit has been simulated using PSpice, the Capture screen, shown in Figure 2.11, displays the node voltages in reverse video. The node voltages are

$$v_1 = 12.00 \text{ V}, \quad v_2 = 1.604 \text{ V}, \quad v_3 = 73.73 \text{ mV} \quad \text{and} \quad v_4 = 6.940 \text{ V}$$

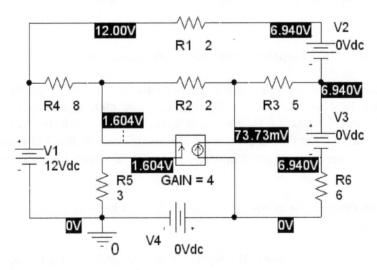

Figure 2.11 The Capture screen shows the node voltages in reverse video after the PSpice simulation is performed.

Step 5 Verify that the simulation results are correct.

Substituting the node voltages from PSpice into the node voltages gives

$$23\,(1.604)-12\,(0.07373)=36.007\approx36$$
$$-55\,(1.604)+21\,(0.07373)-6\,(6.940)=-128.312\neq0$$
$$-6\,(0.07373)+26\,(6.940)=179.998\approx180$$

The node voltages do not satisfy the second node equation. Something is wrong. Similarly substituting the mesh currents from PSpice into the mesh equations gives

$$17\,(2.530)-8\,(3.829)-2\,(3.295)-5\,(1.157)=-0.047\approx0$$
$$-8\,(2.530)+11\,(3.829)-3\,(3.295)=11.994\approx12$$
$$-7\,(2.530)-3\,(3.829)+5\,(3.295)+11\,(1.157)=0.115\neq0$$
$$-4\,(3.829)+3\,(3.295)+(1.157)=-4.264\neq0$$

The mesh currents do not satisfy the third and fourth mesh equations.

The simulation results are not correct. Perhaps the circuit analyzed by PSpice was different from the circuit represented by the mesh and node equations. Comparing the circuits in Figures 2.8 and 2.9 shows that indeed that is the case. In Figure 2.8, i_a, the controlling current of the CCCS is directed downward through the 3 Ω resistor while in Figure 2.9, that current is directed upward. Figure 2.12 shows the corrected circuit in Capture.

Run PSpice to simulate the corrected circuit. Figure 2.13 shows the new PSpice output. The mesh currents are

$$i_1 = 4.160 \text{ A}, \quad i_2 = 6.381 \text{ A}, \quad i_3 = 8.304 \text{ A} \quad \text{and} \quad i_4 = 0.6133 \text{ A}$$

After simulation the circuit using PSpice, the Capture screen, shown in Figure 2.14, displays the node voltages in reverse video. The node voltages are

$$v_1 = 12.00 \text{ V}, \quad v_2 = -5.768 \text{ V}, \quad v_3 = -14.06 \text{ V} \quad \text{and} \quad v_4 = 3.680 \text{ V}$$

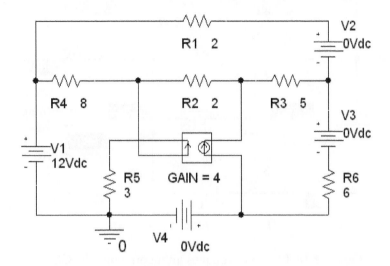

Figure 2.12 The circuit in Capture after correcting the CCCS.

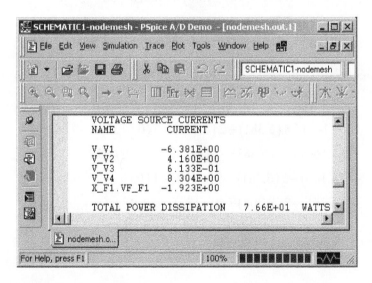

Figure 2.13 The PSpice output file shows the mesh currents.

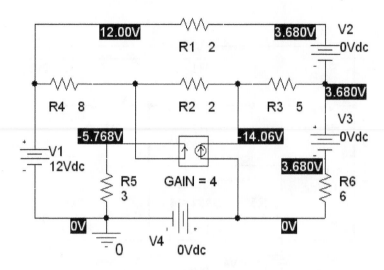

Figure 2.14 The node voltages after correcting the CCCS.

Substituting the node voltages from PSpice into the node equations gives

$$23\left(-5.768\right)-12\left(-14.06\right)=36.056\approx36$$
$$-55\left(-5.768\right)+21\left(-14.06\right)-6\left(3.680\right)=-0.1\approx0$$
$$-6\left(-14.06\right)+26\left(3.680\right)=180.04\approx180$$

The node voltages now do satisfy the node equations. Similarly substituting the mesh currents from PSpice into the mesh equations gives

$$17\left(4.160\right)-8\left(6.381\right)-2\left(8.304\right)-5\left(0.6133\right)=-0.0025\approx0$$
$$-8\left(4.160\right)+11\left(6.381\right)-3\left(8.304\right)=11.999\approx12$$
$$-7\left(4.160\right)-3\left(6.381\right)+5\left(8.304\right)+11\left(0.6133\right)=0.0033\approx0$$
$$-4\left(6.381\right)+3\left(8.304\right)+\left(0.6133\right)=0.0013\approx0$$

The mesh currents now do satisfy the mesh equations.

Step 6 Report the answer to the circuit analysis problem.

The node and mesh equations are correct.

2.3 Comparing DC Circuits

PSpice can be used to compare two circuits. For example, suppose we plan to make a change to a circuit and would like to compare the modified circuit to the original circuit. First, draw the original circuit in the OrCAD workspace. Next copy the circuit drawing by "Cutting and Pasting" the original circuit drawing. Modify the duplicate circuit as desired. Finally, use PSpice to simultaneously analyze the original circuit and the modified circuit. The simulation results provide a comparison of the two circuits.

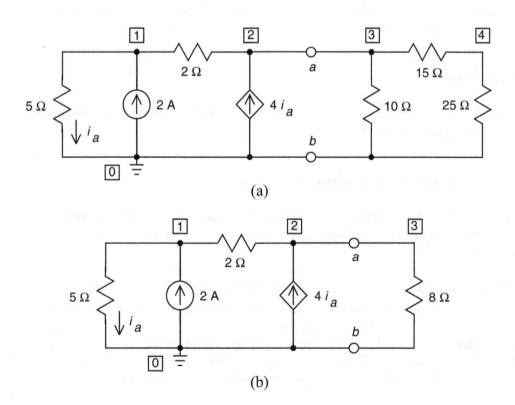

Figure 2.15 (a) A circuit and (b) an equivalent circuit.

Example 2.3 The circuit shown in Figure 2.15a is separated into two parts by the terminals *a-b*. The part of the circuit to the right of terminals consists of three resistors. The 15 Ω resistor is in series with the 25 Ω resistor and the series combination is in parallel with the 10 Ω resistor. In Figure 2.15b, these three resistors have been replaced by a single, equivalent resistor. The 8 Ω resistor in Figure 2.15b is said to be equivalent to the combination of the 10, 15 and 25 Ω resistors in Figure 2.15a because replacing one by the other does not disturb the circuit to the left of the terminals. In particular, the node voltages at nodes 1 and 2 of the circuit in Figure 2.15b will have the same values as the node voltages at nodes 1 and 2 of the circuit in Figure 2.15a. Consequently, the element voltage and current of each element to the left of the terminals in Figure 2.15b will have the same values as the element voltage and current of the corresponding element to the left of the terminals in Figure 2.15a.

Step 1 Formulate a circuit analysis problem.

Compare the circuits in Figure 2.15 by calculating the values of the node voltages at node 1 and node 2 of each circuit.

Step 2 Describe the circuit using OrCAD Capture.

Figure 2.16 shows the circuit from Figure 2.15a as described in Capture. Notice that the positions of the 5 Ω resistor and the current source have been interchanged to make it easier to draw the CCCS.

"Cut and Paste" to copy the circuit. Replace the 10, 15 and 25 Ω resistors to the right of the terminals in the duplicate circuit by a single 8 Ω resistor. If necessary, renumber the parts in the duplicate circuit. (For example, select the label R1 in the duplicate circuit by left-clicking. Right-click to pop up a menu as shown in Figure 1.11. Select "Edit Properties to pop up the "Display Properties" dialog box. Change "R1" to "R6" in the dialog box. Select "OK" to finish.) The result is shown in Figure 2.17.

The circuit shown Figure 2.17 appears to consist of two separate parts, but it is a single connected circuit. The ground symbol marks the bottom nodes of each part as being the same node, so the circuit has only one part.

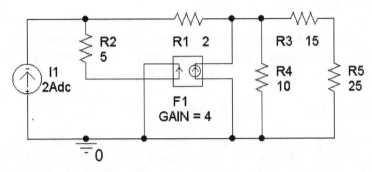

Figure 2.16. The circuit from Figure 2.15*a*, represented using OrCAD Capture.

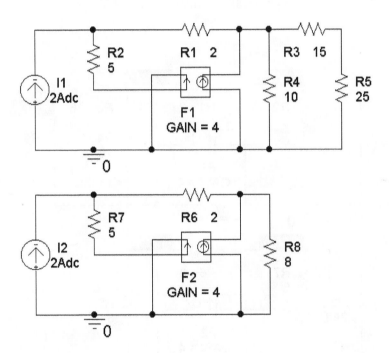

Figure 2.17 Both circuits from Figure 2.15 described in OrCAD Capture.

Step 3 Simulate the circuit using PSpice.

Select **PSpice / New Simulation Profile** to specify the "Analysis type" to be "Bias Point" as described in Chapter 1. Select **PSpice / Run** to run the PSpice simulation.

Step 4 Display the results of the simulation.

After the circuit has been simulated using PSpice, the Capture screen, shown in Figure 2.18, displays the node voltages in reverse video. Comparing Figure 2.18 to Figure 2.15 shows that the values of the node voltages at node 1 and 2 of both circuits in Figure 2.15 are

$$v_1 = -5.882 \text{ V and } v_2 = -12.24 \text{ V}$$

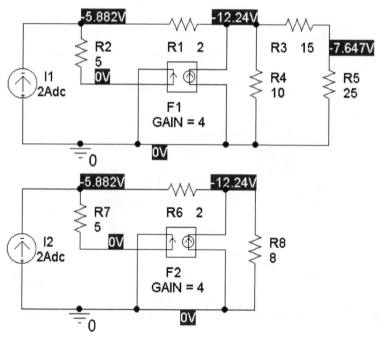

Figure 2.18. The node voltages.

Step 5 Verify the simulation results.

As predicted, the node voltages at nodes 1 and 2 of the circuit in Figure 2.15*b* will have the same values as the node voltages at nodes 1 and 2 of the circuit in Figure 2.15*a*. That's enough to convince us that the simulation is correct.

Step 6 Report the result.

The two circuits in Figure 2.15 are equivalent in the sense that replacing the combination of the 10, 15 and 25 Ω resistors with the 8 Ω resistor did not change the values of the node voltages at nodes 1 and 2 of the circuit.

2.4 Thevenin Equivalent Circuits

Figure 2.19*a* shows a circuit and Figure 2.19*b* shows the Thevenin equivalent circuit. The Thevenin equivalent is determined by finding a value for V_{oc}, the open circuit voltage, and for R_{th}, the Thevenin resistance.

Figure 2.20*a* illustrates a procedure for calculating the value of V_{oc}. An open circuit, represented by a resistor having infinite resistance, is connected to terminals *a* and *b*. The value of the voltage across that open circuit is V_{oc}.

Figure 2.20*b* illustrates a procedure for calculating the value of I_{sc}, called the short circuit current. A short circuit, represented by a 0 V voltage source, is connected to terminals *a* and *b*. The value of the current in that short circuit is I_{sc}. The Thevenin resistance, R_{th}, is calculated from V_{oc} and I_{sc} using

$$R_{th} = \frac{V_{oc}}{I_{sc}}$$

Example 2.4 Use PSpice to calculate the Thevenin equivalent circuit of the circuit shown in Figure 2.19*a*.

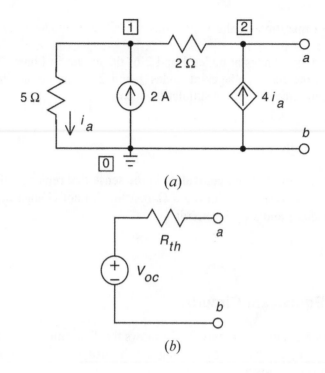

Figure 2.19 (*a*) A circuit and (*b*) its Thevenin equivalent circuit.

Step 1 Formulate a circuit analysis problem.

Find the Thevenin equivalent circuit for the circuit shown in Figure 2.19*a*.

Step 2 Describe the circuit using OrCAD Capture.

Start OrCAD Capture. Create a new project. Place the parts, adjust parameter values and wire the parts together as shown in Figure 2.21.

Make a second copy of the circuit. If necessary, renumber the parts in the duplicate circuit. (For example, select the label R1 in the duplicate circuit by left-clicking on the label. Right-click to pop up a menu as shown in Figure 1.11.

Select "Edit Properties" to pop up the "Display Properties" dialog box. Change "R1" to "R3"in the dialog box. Select "OK" to finish.) The result is shown in Figure 2.22.

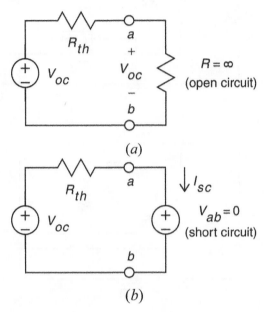

(a)

(b)

Figure 2.20 Calculating (a) V_{oc} and (b) I_{sc}.

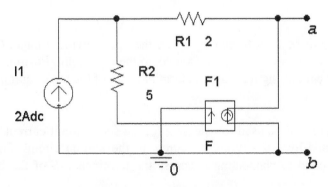

Figure 2.21 The circuit from Figure 2.19a, represented using OrCAD Capture.

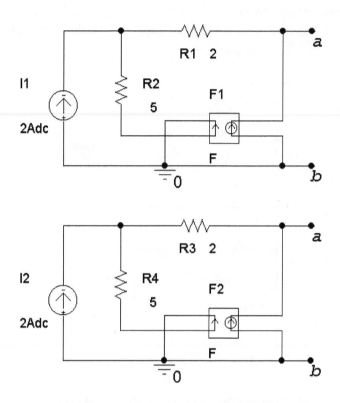

Figure 2.22 Two copies of the circuit from Figure 2.21.

The top circuit will be used to calculate V_{oc}, the open circuit voltage. Connect an open circuit, represented by a very large resistance, to the top circuit. Position this open circuit to correspond to connecting it across the terminals *a-b* of the top circuit in Figure 2.22.

The bottom circuit will be used to calculate I_{sc}, the short circuit current. Connect a short circuit, represented by a 0 V voltage source, to the bottom circuit. Position this short circuit to correspond to connecting it across the terminals *a-b* of the bottom circuit in Figure 2.22.

Figure 2.23 shows the circuits that will be used to calculate V_{oc} and I_{sc}.

Step 3 **Simulate** the circuit using PSpice.

Select **PSpice / New Simulation Profile** from the OrCAD Capture menus to pop up the "New Simulation" dialog box. Specify a simulation name, then select "Create" to close "New Simulation" dialog box and pop up the "Simulation Settings" dialog box. In the "Simulation Settings" dialog box, select "Bias Point" as the analysis type. Finally, select "OK" to close the "Simulation Settings" dialog box and return to the Capture screen.

Select **PSpice / Run** from the OrCAD Capture menus to run the simulation.

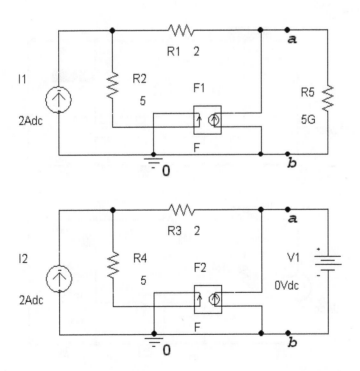

Figure 2.23 Connect (*a*) an open circuit and (*b*) a short circuit across the terminals.

Step 4 Display the results of the simulation, for example, using Probe.

After simulation the circuit using PSpice, the Capture screen, shown in Figure 2.24, displays the node voltages in reverse video. The voltage across R5 is V_{oc} so

$$V_{oc} = -8.667 \text{ V}$$

Figure 2.25 shows the PSpice output file. The current in the voltage source is I_{sc} so

$$I_{sc} = 3.714 \text{ A}$$

The Thevenin resistance is

$$R_{th} = \frac{V_{oc}}{I_{sc}} = \frac{-8.667}{3.714} = -2.334 \ \Omega$$

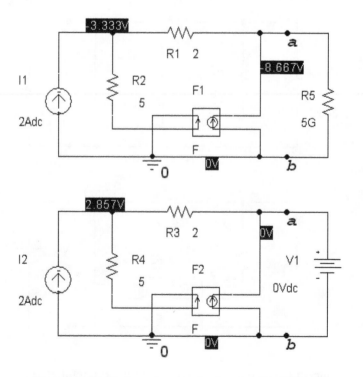

Figure 2.24 Node voltage calculated using PSpice.

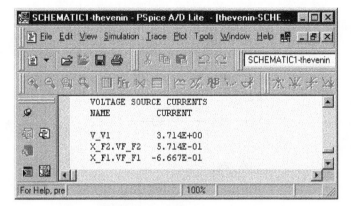

Figure 2.25 The PSpice output file.

Step 5   Verify that the simulation results are correct.

In this example we will use a second PSpice simulation to check the values of R_{th}, and V_{oc}. This second simulation is based on the circuits shown in Figure 2.26.

In Figure 2.26a, a resistor with resistance equal to R_{th}, is connected between terminals a and b of the Thevenin equivalent circuit. The value of the voltage across this resistor should be equal to $V_{oc} / 2$. In Figure 2.26b, a voltage source with voltage equal to V_{oc} is connected between terminals a and b of the Thevenin equivalent circuit. The value of the current in this voltage source should be equal to zero.

Figure 2.27 shows a modification of the circuits simulated using PSpice. In the top circuit, the value of the resistance connected between terminals a and b has been changed to be equal to the value of R_{th}. PSpice calculated the value of the voltage across this resistor to be $-4.334 \text{ V} = V_{oc} / 2$.

In the bottom circuit, the value of the voltage of the voltage source connected between terminals a and b has been changed to be equal to the value of V_{oc}. The PSpice output file shown in Figure 2.28 shows that the current in the voltage source is -0.0001429 A, approximately zero.

These are the expected results. The results of the new simulation agree with the results of the original simulation. The simulation results are correct.

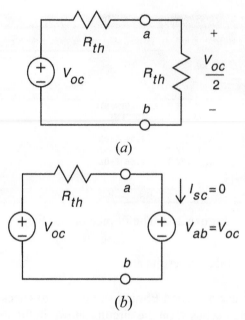

$$(a)$$

$$(b)$$

Figure 2.26 Tests of the values of (*a*) R_{th}, and (*b*) V_{oc}.

Step 6 Report the answer to the circuit analysis problem.

The Thevenin equivalent circuit is shown in Figure 2.19*b*. The values of the open circuit voltage and the Thevenin resistance are

$$V_{oc} = -8.668 \text{ V} \quad \text{and} \quad R_{th.} = -2.334 \text{ } \Omega.$$

55

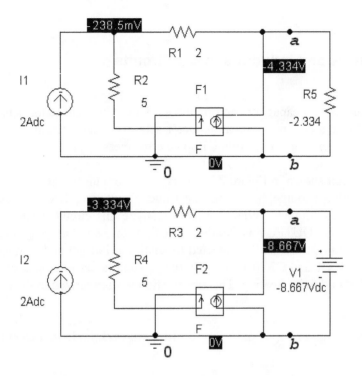

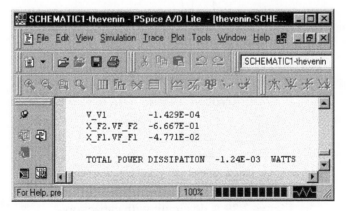

Figure 2.27 Testing the values of R_{th}, and V_{oc}.

Figure 2.28 The PSpice output file.

2.5 Capacitors and Inductors in DC Circuits

DC circuits can contain capacitors and inductors. However a capacitor acts like an open circuit when it is in a dc circuit and an inductor acts like a short circuit when it is in a dc circuit. These facts explain some cryptic PSpice error messages.

Consider the circuit shown in Figure 2.29. This is a dc circuit because the only input, the voltage of the voltage source, has a constant value. The node between capacitors C1 and C2 has been named "output." When asked to analyze this circuit, PSpice produces the error message "Node OUTPUT is floating." At first, this error message is confusing. A floating node is a node that isn't connected to anything, but the node named "output" is connected to both C1 and C2. It is because the capacitors act like open circuits that the node "output" is said to be floating. This observation is summarized in the PSpice rule:

There must be a dc path to ground from each node in a circuit.

Capacitors are ignored when looking for a "dc path."

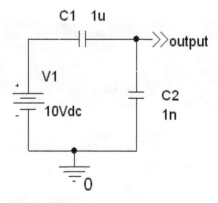

Figure 2.29 A dc circuit with capacitors.

57

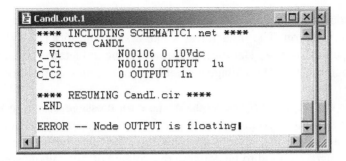

Figure 2.30 A PSpice error message.

Ironically, the error message is eliminated by connecting a very large resistance between the node "output" and the ground node, as shown in Figure 2.31. The resistor provides the required "dc path" from the node "output" to ground. Because the value of the resistance is so large, the resistance acts like an open circuit. Adding an open circuit in this way will not disturb the currents or voltages of the original circuit. (The circuit in Figure 2.29 does not have a unique solution. The node voltages shown in Figure 2.31 represent one solution, but not the only solution, the circuit in Figure 2.29.)

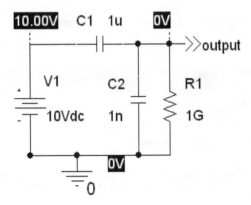

Figure 2.31 Eliminating the PSpice error message.

The circuit shown in Figure 2.32 is a dc circuit containing an inductor. This circuit presents a contradiction. The inductor acts like a short circuit. The value of the voltage across a short circuit is zero. But the voltage across the inductor is equal to the voltage of the voltage source, 10 V. The voltage cannot have two values, both 0 and 10 V.

Analyzing this circuit with PSpice produces the error message shown in Figure 2.33. This error message suggests a solution to the problem. A very small resistor will be inserted into the loop as shown in Figure 3.34. This resolves the contradiction, as the inductor voltage is no longer required to be equal to the voltage of the voltage source. Because the value of the resistance is so small, the resistance acts like a short circuit. Adding a short circuit in this way will not disturb the currents or voltages of the original circuit.

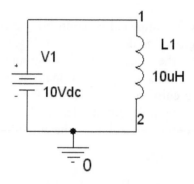

Figure 2.32 A dc circuit with an inductor.

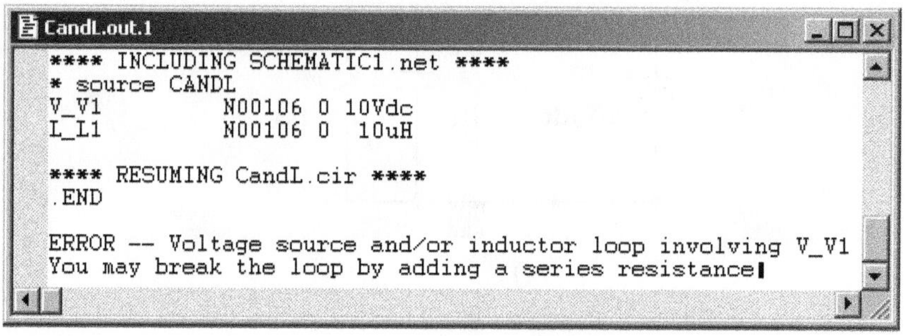

```
CandL.out.1                                             _ □ ×
**** INCLUDING SCHEMATIC1.net ****
* source CANDL
V_V1          N00106 0 10Vdc
L_L1          N00106 0  10uH

**** RESUMING CandL.cir ****
.END

ERROR -- Voltage source and/or inductor loop involving V_V1
You may break the loop by adding a series resistance
```

Figure 2.33 A PSpice error message.

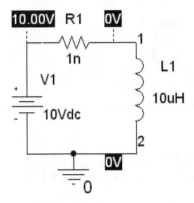

Figure 2.34 Eliminating the PSpice error message.

2.6 Summary

A dc circuit is a circuit in which the voltages of all independent voltage sources and the currents of all independent current sources have constant values. PSpice can analyze a dc circuit to determine the values of the node voltages and also the values of the currents in voltage sources

In this chapter we considered four examples. The first example illustrated analysis of circuits containing dependent sources. The second illustrated the use of PSpice to check the node or mesh equations of a circuit to verify that these equations are correct. The third example used PSpice to compare two dc circuits. The fourth example illustrated using PSpice to find the Thevenin equivalent circuit of a given circuit.

Lastly, we considered some PSpice error messages and their remedies.

CHAPTER 3

Variable DC Circuits

A "variable dc circuit" is a dc circuit in which the constant input is allowed to change. The input to a dc circuit has a constant value and so the values of all currents and voltages of the circuit, including mesh currents and node voltages, are constants. Consideration of variable dc circuits starts by asking the question "What would happen if the value of the input was changed from one constant to another, different constant?" Once the notion of changing the value of the constant input is considered, we can imagine varying the value of the input over a range of constant values.

One can imagine a laboratory experiment in which the value of the input to a circuit is set to a particular constant value and the resulting (constant) value of the circuit output is measured. Next, the value of the input is changed to a new constant value and the new value of the output is measured. This experiment is repeated for a range of constant input values. After all the data has been recorded, a plot can be constructed to display the relationship between the output of the circuit and the input.

PSpice simulates this type of experiment with its "DC Sweep" analysis.

3.1 DC Sweep

The circuit shown in Figure 3.1 is called a voltage divider. The output voltage v_o is related to the input voltage v_i by the equation

$$v_o = \frac{R_2}{R_1 + R_2} v_i \tag{3.1}$$

The gain of the voltage divider is

$$gain = \frac{v_o}{v_i} = \frac{R_2}{R_1 + R_2} \tag{3.2}$$

Plotting v_o as a function of v_i produces a straight line that passes through the origin. The value of the slope of this line is equal to the value of the gain of the voltage divider. Let's use PSpice to analyze the voltage divider.

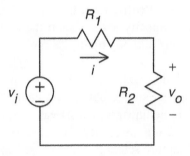

Figure 3.1 A voltage divider.

Example 3.1 Use PSpice to analyze the voltage divider shown in Figure 3.1.
Step 1 Formulate a circuit analysis problem.
The resistances in Figure 3.1 are represented symbolically, by R_1 and R_2, rather than numerically, by values such as 4 Ω and 8 Ω. PSpice is able to analyze circuits numerically rather than symbolically. In order to analyze the voltage divider using PSpice, it is necessary to provide values for R_1 and R_2.
Let $R_1 = 4$ Ω and $R_2 = 8$ Ω. Plot the output voltage of the voltage divider, v_o, as a function of the input voltage, v_i.

Step 2 Describe the circuit using OrCAD Capture.
Start OrCAD Capture. Create a new project. Place the parts, adjust parameter values and wire the parts together as shown in Figure 3.2.

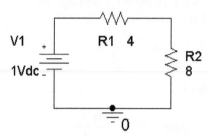

Figure 3.2 The voltage divider in OrCAD Capture.

Step 3 Simulate the circuit using PSpice.

Select **PSpice / New Simulation Profile** from the OrCAD Capture menus to pop up the "New Simulation" dialog box. Specify a simulation name, then select "Create" to close the "New Simulation" dialog box and pop up the "Simulation Settings" dialog box shown in Figure 3.3.

In the "Simulation Settings" dialog box, select "DC Sweep" as the analysis type. Specify the sweep variable to be a voltage source having the name "V1" as shown in Figure 3.3. Next, specify the sweep type to be linear with start value equal to –10, stop value equal to 10 and an increment equal to 1. Finally, select "OK" to close the "Simulation Settings" dialog box and return to the Capture screen.

Select **PSpice / Run** from the OrCAD Capture menus to run the simulation.

Step 4 Display the results of the simulation, for example, using Probe.

After a successful DC Sweep simulation, Probe will open automatically in a Schematics window, as shown in Figure 3.4. Select **Trace / Add Trace** to pop up the "Add Traces" dialog box shown in Figure 3.5. Select V(R2:2) from the list of "Simulation Output Variables," then "OK" to see the plot shown in Figure 3.6. (Probe displays plots using white characters on a black background. This works well on the computer screen, but less well on paper. The plots will be displayed using black characters on a white background in this manual.)

Step 5 Verify that the simulation results are correct.

Figure 3.6 shows the plot of v_o or V(R2:2), as a function of v_i or V_V1. This plot consists of a straight line that goes through the origin and also through the point (3,2). The value of slope of this line is 2/3. When $R_1 = 4\ \Omega$ and $R_2 = 8\ \Omega$, the value of the gain of the voltage divider is also 2/3. The value of the slope of this line is equal to the value of the gain of the voltage divider, as expected. The simulation results are correct.

Probe provides tools to make the plot easier to read. As an example, we will use Probe to label some points on the straight line plot in Figure 3.6. First, we'll remove the grid in Figure 3.6 to make room for the new labels. Select **Plot / Axis Settings** from the Schematics menus to pop up an "Axis Settings" dialog box. Next, remove the grids from both the x and y axis. (For example, select the "X Grid" tab, then select "None" for both the major and minor grids.) Select "OK" to return to the Schematics window. Select **Trace / Cursor / Display** from the Schematics menus to activate the cursor. (This is a toggle, so selecting **Trace / Cursor / Display** will also deactivate an active cursor.) The cursor is positioned using the arrow keys or by left clicking on a trace using the mouse. When the cursor is positioned as desired, select **Plot / Label / Mark** from the Schematics menus to mark the point. Deactivate the cursor by selecting **Trace / Cursor / Display** from the Schematics menus. Then position the label as desired using the mouse. Repeat this procedure to label a second point on the line. The labeled plot is shown in Figure 3.7. The label (3.0303, 2.0202) indicates that V(R2:2) = 2.0202 when V1_1 = 3.0303. The slope of the line can be calculated from the labeled points and is seen to be 2/3.

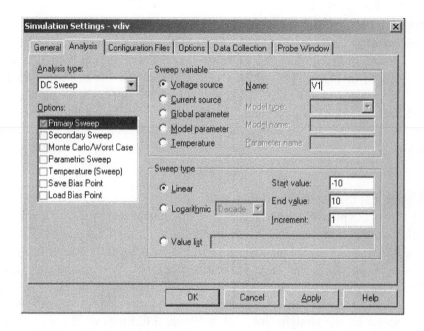

Figure 3.3 Setting up a DC Sweep Simulation.

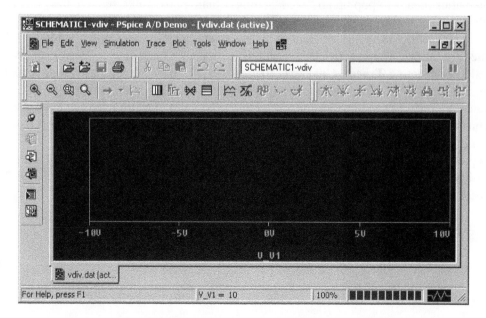

Figure 3.4 Probe starts automatically after a successful PSpice simulation.

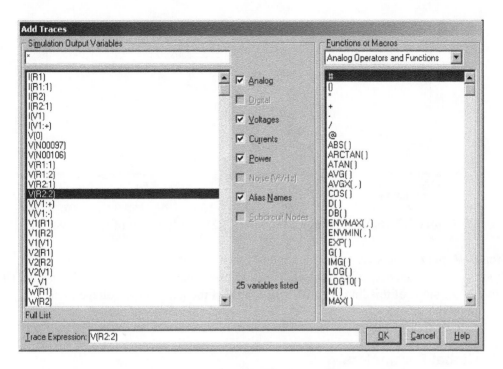

Figure 3.5 The "Add Traces" dialog box.

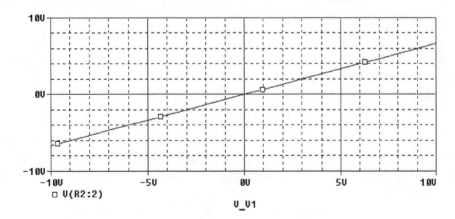

Figure 3.6 The plot of v_o = V(R2:2) as a function of v_i = V_V1.

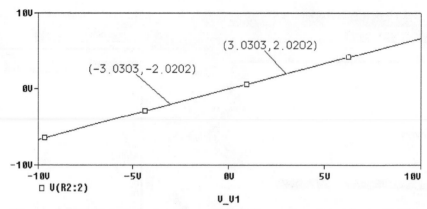

Figure 3.7 The labeled plot of v_o = V(R2:2) as a function of v_i = V_V1.

Step 6 Report the answer to the circuit analysis problem.
Plotting v_o as a function of v_i produces a straight line that passes through the origin. The value of the slope of this line is equal to the value of the gain of the voltage divider.

3.2 Global Parameters

Consider again the voltage divider shown in Figure 3.1. The input resistance of this voltage divider is

$$R_i = R_1 + R_2 \tag{3.3}$$

The input resistance of the voltage divider is significant because it determines the value of the power, p_i, that the voltage source must supply.

$$p_i = \frac{v_i^2}{R_1 + R_2} \tag{3.4}$$

It is reasonable to specify the voltage divider by giving required values of input resistance and gain and to design the voltage divider by determining values of R_1 and R_2 that produce the specified input resistance and gain. Solving equations 3.2 and 3.3 for R_1 and R_2 gives

$$R_1 = (1 - gain) \times R_i \tag{3.5}$$

and

$$R_2 = gain \times R_i \tag{3.6}$$

A PSpice feature called "Global Parameters" makes it possible to specify values of the input resistance and gain and then use these values to calculate corresponding values for R_1 and R_2.

Example 3.2 Use PSpice to analyze the voltage divider shown in Figure 3.1.

Step 1 Formulate a circuit analysis problem.

Design a voltage divider having an input resistance equal to 20 Ω and a variable gain. Plot the power supplied by the voltage source and the gain for several voltage dividers having different gains.

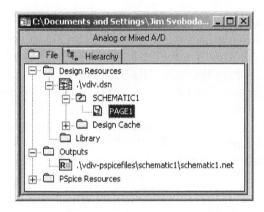

Figure 3.8 Reopening a saved project.

Step 2 Describe the circuit using OrCAD Capture.

Start OrCAD Capture. Open the voltage divider project from Example 3.1. OrCAD Capture will open the window shown in Figure 3.8. Double-click on PAGE1. The voltage divider will appear in an OrCAD Capture workspace as shown in Figure 3.2.

First, we will add two global parameters to represent the input resistance and gain. Next, we will modify the properties of the resistors R_1 and R_2 so that their resistances depend on the global parameters.

Select **Place / Parts** from the OrCAD Capture menus to open the "Place Parts" dialog box. Global parameters are parts named "PARAM" that are contained in the "SPECIAL" library. (It may be necessary to add this library. The file is called special.olb and resides in the "PSpice" folder.) Place a "PARAM" part in the OrCAD Capture workspace as shown in Figure 3.9.

Next select the PARAM part and edit its properties. (Left-click on the PARAM part to select it, then right-click to pop up a menu. Select "Edit Properties…" from that menu.) A property sheet will open as shown in Figure 3.10. Select "New Column." A new Column dialog box will appear. Supply the name "Ri" and the value 20. Click "OK" to close the dialog box. The PARAM part now has a property called "Ri" having a value 20 as shown in Figure 3.11.

To display the global parameter and its value left-click on the label "Ri" above the value 20, then left-click the button labeled "Display." Select "Name and Value" in the dialog box that pops-up, then left-click the "OK" button to close that dialog box.

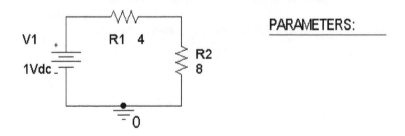

Figure 3.9 The voltage divider after adding a PARAM part.

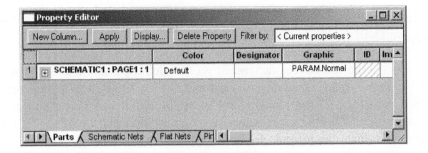

Figure 3.10 The Property Editor.

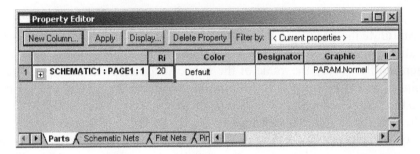

Figure 3.11 The global parameter "Ri" has a value of 20.

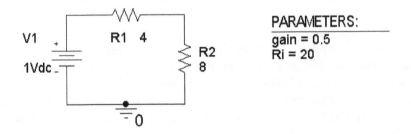

Figure 3.12 The global parameters.

Repeat this procedure to give the other global parameter the name "gain" and the value 0.5. The circuit will appear as shown in Figure 3.12.

Now that the global parameters have been defined, we will edit the properties of the resistors to make the resistances dependent on the global parameters. Select resistor R2 and edit its properties. (Left-click on resistor R2, then right-click to pop up a menu. Select "Edit Properties…" from that menu.) Scroll through the properties to find "Value." Change the value of "Value" from 8 to {Ri*gain} as shown in Figure 3.13. The braces tell PSpice to evaluate the formula contained within the braces to determine the value of the resistance. Edit the properties of R1 to change the value from 4 to {Ri*(1-gain)}. Now the circuit will appear as shown in Figure 3.14.

Step 3   Simulate the circuit using PSpice.

A simulation profile for this project was defined in Example 3.1. We need to edit this simulation profile to vary the value of the gain in addition to sweeping the value of the input voltage. Select **PSpice / Edit Simulation Profile** from the OrCAD Capture menus to pop up the " Simulation Settings" dialog box as shown in Figure 3.3.

Check the box for "Parametric Sweep" in the list of Options as shown in Figure 3.15. Select Global Parameter as the Sweep variable. Provide the Parameter name "gain" and indicate a start value of 0.1, a stop value of 0.9 and an increment of 0.2. (These values provide a variety of gains while avoiding setting the resistance of either resistor to zero. A zero resistance would cause a divide by zero error.)

Select **PSpice / Run** from the OrCAD Capture menus to run the simulation.

Step 4   Display the results of the simulation, for example, using Probe.

After a successful DC Sweep simulation, Probe, the graphical post-processor, will open automatically in a Schematics window. An "Available Sections" dialog box will also pop up as a result of the parametric sweep. The "Available Sections" dialog box lists the five values, 0.1, 0.3, 0.5, 0.7 and 0.9, of the global parameter "gain" used in this simulation. Click the button "OK" to close the "Available Sections" dialog box. The Schematics window is shown in Figure 3.4.

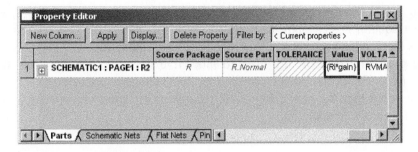

Figure 3.13 Editing the properties of R2.

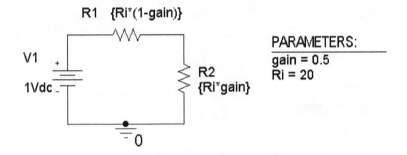

Figure 3.14 The voltage divider using global parameters.

The simulation results will be displayed using two plots, one for the power supplied by the voltage source and one for the gain of the voltage divider. Each plot will display five traces, corresponding to the five values of the gain.

Select **Plot / Add Plot to Window** to add a second plot. Two empty plots will appear, one above the other. Select the top plot by left-clicking the mouse on the top plot.

Select **Trace / Add Trace** to pop up the "Add Traces" dialog box shown in Figure 3.5 Select W(V1) from the list of "Simulation Output Variables." W(V1) is the power corresponding to the voltage source. PSpice uses the passive convention for all elements, including sources, so W(V1) is the power delivered to the voltage source, rather than the power supplied by the voltage source. Edit the Trace Expression to be "–W(V1)" then click the "OK" button to see the plots shown in Figure 3.16.

Notice the label for the top plot. The five geometric symbols correspond to the five values of the global parameter "gain" used in this simulation. The order of these symbols corresponds to the order of values of "gain." Recall the gain started at a value of 0.1 and increased by increments of 0.2 until it reached 0.9. Consequently, the first geometric symbol, a square, corresponds to the first value of gain, 0.1. The second geometric symbol, a diamond, corresponds to the second value of gain, 0.3.

The five geometric symbols are used to label five traces. In this case the five traces coincide. (Recall that the power supplied by the voltage source depends on the input resistance rather than the gain. The global parameter Ri remained constant at a value of 20 for all five simulations.)

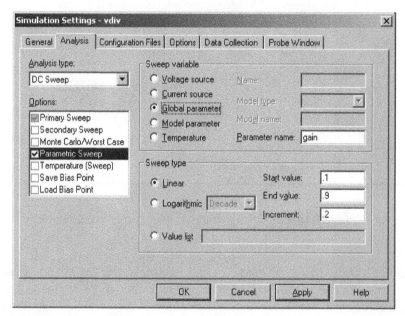

Figure 3.15 Edit the simulation settings to vary the gain of the voltage divider.

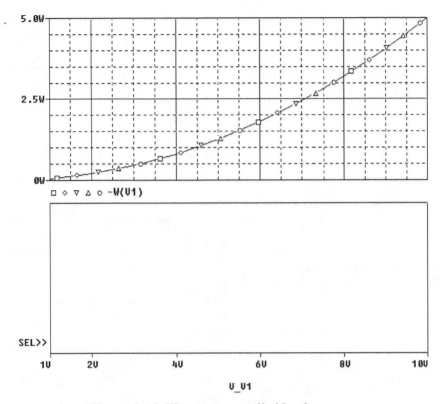

Figure 3.16 The power supplied by the source.

Select the bottom plot by left-clicking the mouse on the bottom plot. Select **Trace / Add Trace …** from the Schematics menus to pop up the dialog box shown in Figure 3.5. Select V(R2:2) from the list of "Simulation Output Variables." V(R2:2) is the output voltage, v_o. Next select V_V1 from the list of "Simulation Output Variables." V_V1 is the input voltage, v_i. At this point the Trace Expression is "V(R2:2) V_V1." Edit the Trace Expression to be "V(R2:2)/V_V1," corresponding to the gain, v_o / v_i. Left-click the "OK" button to see the plot shown in Figure 3.17.

Figure 3.17 shows unexpected behavior when v_i is zero. Both the input and output voltages are zero when the input voltage is zero, so the gain is undefined. It is appropriate to exclude this case. Select **Plot / Axis Settings…** from the Schematics menu. Choose the "X Axis" tab. Select a user defined data range and specify a starting voltage of 1.0V. Now the plot appears as shown in Figure 3.18.

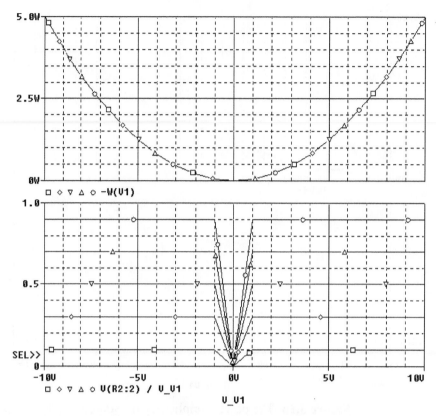

Figure 3.17 Plots of power and gain.

Step 5   Verify that the simulation results are correct.

The top plot shows that the power supplied by the voltage source does not depend on the gain of the voltage divider. Further, this plot shows that approximately 4.5 W are supplied when the input is 9.5 V. This agrees with a calculated value of

$$\frac{9.5^2}{20} = 4.5125 \;\; \text{W}$$

The bottom plot shows that the ratio of the output voltage to the input voltage agrees with the specified value of the gain.

The simulation results are correct.

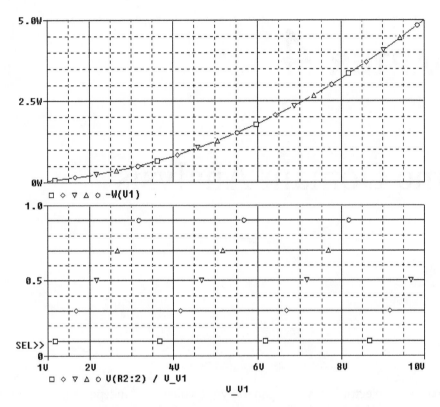

Figure 3.18 Power and gain plots for the voltage divider.

Step 6 Report the answer to the circuit analysis problem.
The voltage divider shown in Figure 3.12 has an input resistance equal to 20 Ω and a variable gain. The power supplied by the voltage source and the gain of several voltage dividers are plotted in Figure 3.18.

3.3 Summary

DC Sweep analysis is used to simulate variable dc circuits, dc circuits in which the constant input is allowed to vary over a range of constant values.

CHAPTER 4

Time Domain Analysis

The inputs to an electric circuit are generally the voltages of independent voltage sources and the currents of independent current sources. PSpice provides a set of voltage and current sources that represent time varying inputs. "Time Domain (Transient)" analysis using PSpice simulates the response of a circuit to a time varying input.

4.1 The Response of an RC Circuit to a Pulse Input

Time domain analysis is most interesting for circuits that contain capacitors or inductors or both. PSpice provides parts representing capacitors and inductors in the "ANALOG" parts library. The part name for the capacitor is "C." The part properties that are of the most interest are the capacitance and the initial condition, both of which are specified using the OrCAD Capture property editor. (The initial condition of a capacitor is the value of the capacitor voltage at time $t = 0$.)

The part name for the inductor is "L." The inductance and the initial condition of the inductor are specified using the property editor. (The initial condition of an inductor is the value of the inductor current at time $t = 0$.)

The voltage and current sources that represent time varying inputs are provided in the "SOURCE" parts library. Table 4.1 summarizes these voltage sources. The voltage waveform describes the shape of the voltage source voltage as a function of time. Each voltage waveform is described using a series of parameters. For example, the voltage of an exponential source, VEXP, is described using $v1$, $v2$, $td1$, $td2$, $tc1$ and $tc2$. The parameters of the voltage sources in Table 4.1 are specified using the property editor.

Table 4.1 PSpice Voltage Sources for Transient Response Simulations

Name	Symbol	Voltage Waveform
VEXP	V1 = V2 = TD1 = TC1 = TD2 = TC2 =	
VPULSE	V1 = V2 = TD = TR = TF = PW = PER =	

Table 4.1 (continued)

Name	Symbol	Voltage Waveform
VPWL	V?	*(waveform showing piecewise linear segments through points $t1, v1$; $t2, v2$; $t3, v3$; $t4, v4$)*
VSIN	VOFF = VAMPL = FREQ = V?	*(damped sinusoid waveform with $vo + va$, vo, df, td, $\frac{1}{freq}$)*

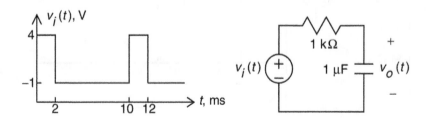

Figure 4.1 An RC circuit with a pulse input.

Example 4.1 The input to the circuit shown in Figure 4.1 is the voltage source voltage, $v_i(t)$. The output, or response, of the circuit is the voltage across the capacitor, $v_o(t)$. The output voltage can be calculated to be

$$v_o(t) = \begin{cases} 4\left(1-e^{-1000t}\right) & 0 < t < 2 \text{ ms} \\ -1 + 4.46\, e^{-1000(t-0.002)} & 2 \text{ ms} < t < 10 \text{ ms} \end{cases} \qquad (4.1)$$

Use PSpice to simulate the response of this circuit to the pulse input shown in Figure 4.1.

Step 1 Formulate a circuit analysis problem.
Plot the voltages $v_i(t)$ and $v_o(t)$ versus time t for the circuit shown in Figure 4.1.

Step 2 Describe the circuit using OrCAD Capture.
Start OrCAD Capture. Create a new project. Place the parts, adjust parameter values and wire the parts together as shown in Figure 4.2. The voltage source is a VPULSE part (see the second row of Table 4.1).

Figure 4.1 shows $v_i(t)$ making the transition from −1 V to 4 V instantaneously. Zero is not an acceptable value for the parameters *tr* or *tf*. Choosing very small value for *tr* and *tf* will make the transitions appear to be instantaneous when using a time scale that shows a period of the input waveform. In this example, the period of the input waveform is 10 ms so 1 ns is a reasonable choice for the values of *tr* and *tf*.

It's convenient to set *td*, the delay before the periodic part of the waveform, to zero. Then the values of *v*1 and *v*2 are −1 and 4, respectively. The value of *pw* is the length of time that $v_i(t) = v2 = 4$ V, so *pw* = 2 ms in this example. The pulse input is a periodic function of time. The value of *per* is the period of the pulse function, 10 ms.

Figure 4.3 shows specification of the voltage source parameters using the OrCAD property editor. (Figure 4.3 shows only the parameters of interest by "hiding" the other parameters. The parameters of interest are found by scrolling through the list of parameters.)

The circuit shown in Figure 4.1 does not have a ground node. PSpice requires that all circuits have a ground node, so it is necessary to select a ground node. Figure 4.2 shows that the bottom node has been selected to be the ground node.

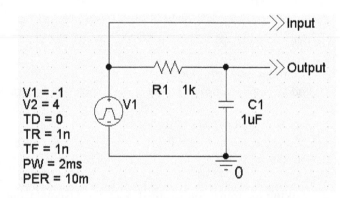

Figure 4.2 The circuit as described in OrCAD Capture.

	PER	PW	TD	TF	TR	V1	V2	Value
1 SCHEMATIC1 : PAGE1 : V1	10m	2ms	0	1ns	1ns	-1	4	VPULSE

Figure 4.3 Specifying the VPULSE source using the property editor.

Step 3 Simulate the circuit using PSpice.

Select **PSpice / New Simulation Profile** from the OrCAD Capture menus to pop up the "New Simulation" dialog box. Specify a simulation name, then select "Create" to close "New Simulation" dialog box and pop up the "Simulation Settings." In the "Simulation Settings" dialog box, select "Time Domain(Transient)" as the analysis type. The simulation starts at time zero and ends at the "Run to time." Specify the "Run to time" as 20 ms to run the simulation for two full periods of the input waveform. Check the "Skip the initial transient bias point calculation (SKIPBP)" checkbox. Finally, select "OK" to close the "Simulation Settings" dialog box and return to the Capture screen.

Select **PSpice / Run** from the OrCAD Capture menus to run the simulation.

Step 4 **Display** the results of the simulation, for example, using Probe.

After a successful Time Domain(Transient) simulation, Probe, the graphical post-processor for PSpice, will open automatically in a Schematics window. Select **Trace /** **Add Trace** to pop up the "Add Traces" dialog box. Add the traces V(OUTPUT) and V(INPUT). Figure 4.4 shows resulting plot after removing the grid and labeling some points.

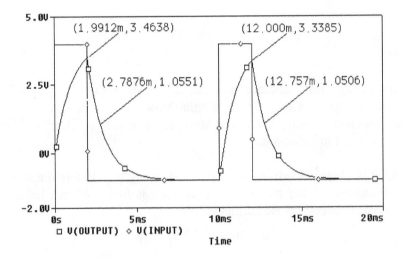

Figure 4.4 The response of the RC circuit to the pulse input.

Step 5 Verify that the simulation results are correct.
Each label in Figure 4.4 indicates the coordinates of a point on the plot of the capacitor voltage, $v_o(t)$. For example, the label (1.9912m, 3.4638) indicates that $v_o(t) = 3.4638$ V when $t = 1.9912$ ms $= 0.0019912$ sec. This value can be checked using Equation 4.1. Substituting t = 1.9912 ms into Equation 4.1 gives $v_o(t) = 3.4539$ V, a difference of 0.3%. Similarly, Figure 4.4 indicates that $v_o(t) = 1.0551$ V when $t = 2.7876$ ms. Substituting t = 2.7876 ms into Equation 4.1 gives $v_o(t) = 1.0290$ V, a difference of 2.4%. The simulation results are correct.

Step 6 Report the answer to the circuit analysis problem.
Plots of the voltages $v_i(t)$ and $v_o(t)$ are shown in Figure 4.4.

4.2 Switches

Time varying voltages and currents can be caused by opening or closing a switch. PSpice provides parts to represent single-pole, single-throw (SPST) switches in the "EVAL" parts library. A voltage controlled switch is available in the "ANALOG" library. These parts are summarized in Table 4.2.

Example 4.2 The circuit shown in Figure 4.5 is at steady state before the switch closes at time $t = 0$. Consequently the current in the inductor is $i_L(t) = 40$ mA before the switch closes. After the switch closes the current is

$$i_L(t) = \left(60 - 20\, e^{-40000\, t}\right) \text{ mA} \qquad (4.2)$$

Use PSpice to simulate the circuit after the switch closes.

Table 4.2 PSpice switch

Symbol	Description	PSpice Name	PSpice Library
TCLOSE = 0 1 ⟶ 2 U?	open switch, will close at t = TCLOSE	Sw_tClose	EVAL
TOPEN = 0 1 ⟶ 2 U?	closed switch, will open at t = TOPEN	Sw_tOpen	EVAL
S? S VOFF = 0.0V VON = 1.0V	voltage controlled switch	S	ANALOG

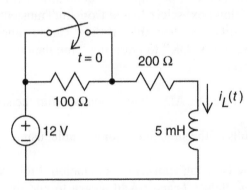

Figure 4.5 A circuit with a switch.

Step 1 Formulate a circuit analysis problem.

After the switch closes, the circuit shown in Figure 4.5 has a time constant with a value of 25 μs. Plot the inductor current, $i_L(t)$, for the first 150 μs (6 time constants) after the switch closes.

Step 2 Describe the circuit using OrCAD Capture.

Start OrCAD Capture. Create a new project. Place the parts, adjust parameter values, and wire the parts together as shown in Figure 4.6.

The circuit shown in Figure 4.5 does not have a ground node. PSpice requires that all circuits have a ground node, so it is necessary to select a ground node. Figure 4.6 shows that the bottom node has been selected to be the ground node.

The initial condition of the inductor in Figure 4.5 is $i_L(0) = 40$ mA. The PSpice part representing an inductor has a property named "IC" that represents the initial current in the inductor. This value of the initial condition can be specified using the OrCAD property editor as shown in Figure 4.7. (Figure 4.7 shows only the parameters of interest by "hiding" the other parameters. The parameters of interest are found by scrolling through the list of parameters.)

Step 3 Simulate the circuit using PSpice.

Select **PSpice / New Simulation Profile** from the OrCAD Capture menus to pop up the "New Simulation" dialog box. Specify a simulation name, then select "Create" to close "New Simulation" dialog box and pop up the "Simulation Settings" dialog box. In the "Simulation Settings" dialog box, select "Time Domain(Transient)" as the analysis type. The simulation starts at time zero and ends at the "Run to time." Specify the "Run to time" as 150 us. Finally, select "OK" to close the "Simulation Settings" dialog box and return to the Capture screen.

Select **PSpice / Run** from the OrCAD Capture menus to run the simulation.

Step 4 Display the results of the simulation, for example, using Probe.

After a successful "Time Domain(Transient)" simulation, Probe will open automatically in a Schematics window. Select **Trace / Add Trace** to add the trace I(L1). Figure 4.8 shows resulting plot after removing the grid and labeling one point.

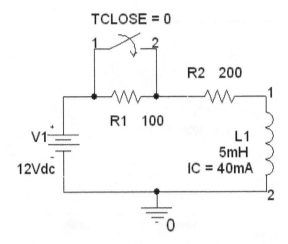

Figure 4.6 The circuit in OrCAD Capture.

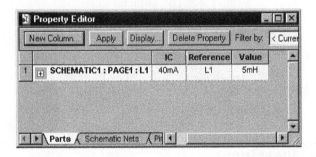

Figure 4.7 Specifying the initial condition.

Step 5 Verify that the simulation results are correct.

Equation 4.2 predicts that the inductor current will be 40 mA when $t = 0$ and also that the inductor current will be 60 mA as t approaches infinity. The plot in Figure 4.8 agrees with both of these predictions.

The plot in Figure 4.8 indicates that $i_L(t) = 53.592$ mA when $t = 30.405$ μs. Substituting $t = 30.405$ μs into Equation 4.2 gives $i_L(t) = 54.073$ mA, a difference of 0.5%. The simulation results are correct.

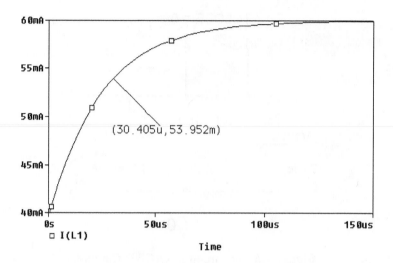

Figure 4.8 The inductor current.

Step 6 Report the answer to the circuit analysis problem.

Figure 4.8 shows inductor current, $i_L(t)$, for the first 6 time constants after the switch in Figure 4.5 closes.

4.3 Piecewise Linear Inputs

The piecewise linear voltage source, shown in row 3 of Table 4-1, makes it possible to incorporate complicated input signals in circuits analyzed using PSpice.

Example 4.3 The voltage of the voltage source in the circuit shown in Figure 4.9 is a piecewise linear function of time. In other words, it consists of a series of straight line segments. Also, consecutive straight line segments share end points. It's convenient to represent these straight line segments by a list of end points. For example, the function $v_i(t)$ can be represented by

$$(0, 0), (2.5, 8), (5, 0), (6, -2), (15, -2), (16, 0), (\infty, 0)$$

Each of the points is represented by an ordered pair, $(t_k, v_k(t_k))$ where t_k is the time in s and $v_k(t_k)$ is the corresponding voltage in V.

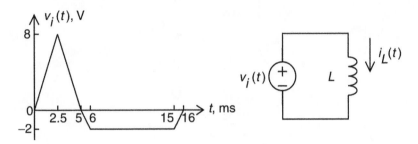

Figure 4.9 A circuit with a piecewise linear input.

Step 1 Formulate a circuit analysis problem.

Simulate the circuit shown in Figure 4.9 to determine the inductor current, $i_L(t)$. Plot the $i_L(t)$ versus t. Assume $i_L(t) = 0$ for $t \le 0$.

Step 2 Describe the circuit using OrCAD Capture.

Start OrCAD Capture. Create a new project. Place the parts, adjust parameter values and wire the parts together as shown in Figure 4.10. The voltage source is a VPWL part (see the third row of Table 4.1).

(R1 has a very small resistance and so acts like a short circuit. PSpice would not analyze the circuit without this resistor, because the circuit includes a loop consisting entirely of voltage sources and inductors. The PSpice error message suggested adding a resistor such as R1.)

The VPWL source is specified using the property editor. (Left-click on the voltage source to select it then right-click to pop up a menu. Select Edit Properties from that menu.) Figure 4.11 shows the property editor being used to represent $v_i(t)$ by the list of points given above.

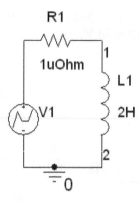

Figure 4.10 The circuit in OrCAD Capture.

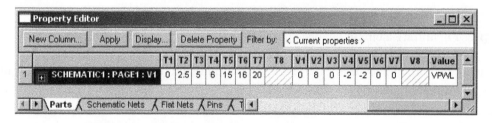

Figure 4.11 Specifying the piecewise linear voltage source.

Step 3 Simulate the circuit using PSpice.

Select **PSpice / New Simulation Profile** from the OrCAD Capture menus to pop up the "New Simulation" dialog box. Specify a simulation name, then select "Create" to close "New Simulation" dialog box and pop up the "Simulation Settings" dialog box. In the "Simulation Settings" dialog box, select "Time Domain(Transient)" as the analysis type. The simulation starts at time zero and ends at the "Run to time." Specify the "Run to time" as 20s to run the simulation past the end of the input waveform. Finally, select "OK" to close the "Simulation Settings" dialog box and return to the Capture screen.

Select **PSpice / Run** from the OrCAD Capture menus to run the simulation.

Step 4 Display the results of the simulation, for example, using Probe.

After a successful Time Domain(Transient) simulation, Probe, the graphical post-processor for PSpice, will open automatically in a Schematics window. Select **Plot / Add Plot to Window** from the menus twice to add two plots. Left-click on the top plot to select it. Then select **Trace / Add Trace** from the menu bar to pop up the "Add Traces" dialog box. Add the trace V(V1:+) to display $v_i(t)$ in the top plot. Display $i_L(t)$ in the middle plot and W(L1), the power delivered to the inductor, in the bottom plot. Figure 4.15 shows resulting plots after removing the grids.

Step 5 Verify that the simulation results are correct.

First, comparing the top plot in Figure 4.12 to the plot in Figure 4.9 shows that the VPWL voltage source represents $v_i(t)$ accurately.

Next, the voltage of the voltage source is equal to the voltage across the inductor so the inductor current, $i_L(t)$, is related to $v_i(t)$ by the equation

$$i_L(t) = \frac{1}{L} \int_0^t v_i(\tau)\, d\tau + i_L(0) = \frac{1}{2} \int_0^t v_i(\tau)\, d\tau$$

This integral can be evaluated using the "area under the curve." The "area under the curve" for the positive part of $v_i(t)$ is 20 V-s so $i_L(t)$ should be 10 A when $t = 5$ s. The "area under the curve" for the negative part of $v_i(t)$ is also 20 V-s so $i_L(t)$ should be 0 A when $t = 16$ s. These observations are consistent with the plot of $i_L(t)$ versus t. The simulation results are correct.

Step 6 Report the answer to the circuit analysis problem.

Figure 4.12 shows the inductor current, $i_L(t)$, plotted versus t, for the circuit in Figure 4.9.

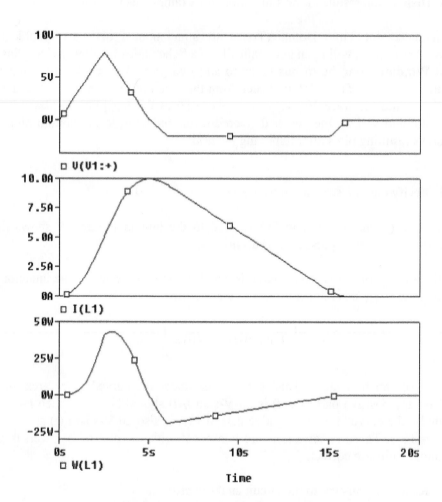

Figure 4.12 The simulation results.

4.4 AC Circuits in the Time Domain

Linear, time-invariant circuits that are excited by sinusoidal inputs and have reached steady state are called AC circuits. AC circuits are usually analyzed in the frequency domain. Chapter 5 describes using PSpice to simulate AC circuits in the frequency domain. In contrast, it is sometimes useful to display the response of an AC circuit in the time domain. PSpice's "Time Domain(Transient)" analysis produces the required time domain responses.

Example 4.4 The circuit shown in Figure 4.13 has been analyzed to determine the capacitor voltage, $v_c(t)$. We will use PSpice to plot the capacitor voltage as a function of time and, also, plot the power delivered to the capacitor as a function of time.

Step 1 Formulate a circuit analysis problem.

Simulate the circuit shown in Figure 4.13 to determine the capacitor voltage, $v_c(t)$. Plot the $v_c(t)$ versus t. Also, plot the power delivered to the capacitor as a function of time, t.

Step 2 Describe the circuit using OrCAD Capture.

Start OrCAD Capture. Create a new project. Place the parts, adjust parameter values and wire the parts together as shown in Figure 4.14. The voltage source is a VSIN part (see the fourth row of Table 4.1). Notice that the voltage source frequency is specified in Hertz rather than rad/s.

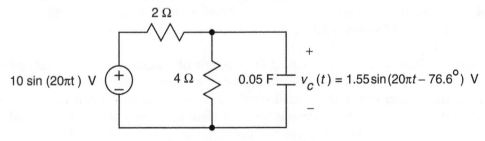

Figure 4.13 An AC circuit.

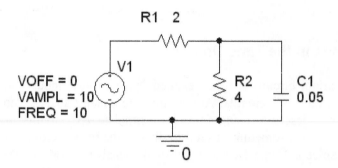

Figure 4.14 The AC circuit in OrCAD Capture.

Step 3 Simulate the circuit using PSpice.

Select **PSpice / New Simulation Profile** from the OrCAD Capture menus to pop up the "New Simulation" dialog box. Specify a simulation name, then select "Create" to close "New Simulation" dialog box and pop up the "Simulation Settings" dialog box. In the "Simulation Settings" dialog box, select "Time Domain(Transient)" as the analysis type. The simulation starts at time zero and ends at the "Run to time." Specify the "Run to time" as 0.8s to run the simulation for eight full periods of the input waveform. Finally, select "OK" to close the "Simulation Settings" dialog box and return to the Capture screen.

Select **PSpice / Run** from the OrCAD Capture menus to run the simulation.

Step 4 Display the results of the simulation, for example, using Probe.
After a successful Time Domain(Transient) simulation, Probe, the graphical post-processor for PSpice, will open automatically in a Schematics window. Select **Trace / Add Trace** from the menus to pop up the "Add Traces" dialog box. Add the trace V(C1:1). Figure 4.15 shows resulting plot after removing the grid.

The plot in Figure 4.15 is disappointing for a couple of reasons. First, it's a very rough representation of a sine function. Second, it takes a while for the capacitor voltage to settle down. In other words, the capacitor voltage includes a transient part as well as the steady state response. In this example we only want the steady state response and so would like to eliminate the transient part of the response.

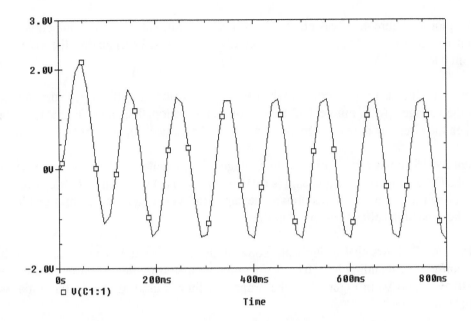

Figure 4.15 A disappointing plot of the capacitor voltage.

Both of these disappointments can be remedied. In order to obtain a smoother plot, select **PSpice / Edit Simulation Profile** from the OrCAD Capture menus to pop up the "Simulation Settings" dialog box. In the "Simulation Settings" dialog box, set the "Maximum step size" to 0.001s. (The period of the sine function is 0.1 second. A maximum step size equal to 0.001 second will cause at least 100 points to be plotted each period, resulting in a smoother plot.) Finally, select "OK" to close the "Simulation Settings" dialog box and return to the Capture screen.

The transient part of the capacitor voltage can be eliminated by adjusting the initial value of the capacitor voltage, $v_c(0)$. From Figure 4.13

$$v_c(t) = 1.55 \sin\left(20\pi t - 76.6°\right) \text{ V}$$

Let $t = 0$, then

$$v_c(0) = 1.55 \sin\left(-76.6°\right) = -1.506 \text{ V}$$

The plot in Figure 4.15 was obtained using the default value of $v_c(0)$, which is zero. The transient part of the capacitor voltage can be eliminated by changing the value of $v_c(0)$ from 0 to -1.506 V.

Some care needs to be exercised when setting the value of $v_c(0)$. Notice that the polarity of the voltage $v_c(t)$ in Figure 4.13 has the plus sign on top. This polarity must also be used when the circuit is drawn in OrCAD Capture if the value $v_c(0) = -1.506$ is to be used.

Figure 4.16 shows the polarity of the voltage of a capacitor as the capacitor is (a) placed in the OrCAD workspace and as it is rotated (b) once, (c) twice and (d) three times. When the circuit in Figure 4.14 was drawn, the capacitor was rotated three times so the polarity of the capacitor voltage has the plus sign on top.

Figure 4.17 shows that a duplicate copy of the AC circuit has been made so that the response corresponding to $v_c(0) = 0$ V can be compared to the response corresponding to $v_c(0) = -1.506$ V. In Figure 4.18 the property editor is used to change the initial voltage of C2 to -1.506 V.

Figure 4.19 shows the voltages of the two capacitors in Figure 4.17. Both of the disappointments with the plots of Figure 4.15 have been remedied. The plots are smoother and the change to the initial capacitor voltage did indeed eliminate the transient part of the capacitor voltage.

Finally, Figure 4.20 shows the power delivered to the capacitor (top plot) and the capacitor voltage. (W(C2), the power delivered to capacitor C2, is one of the functions that PROBE calculated automatically.)

(a)   (b)   (c)   (d)

Figure 4.16 The polarity of the voltage of a capacitor as the capacitor is (a) placed in the OrCAD workspace and as it is rotated (b) once, (c) twice and (d) three times.

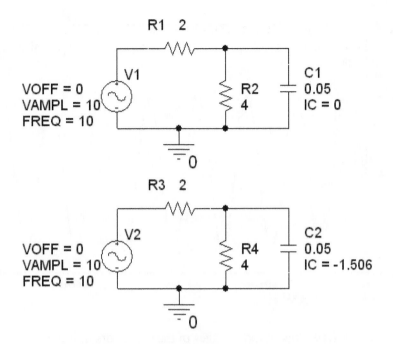

Figure 4.17 Two copies of the AC circuit.

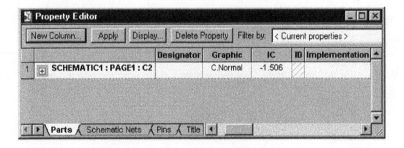

Figure 4.18 Changing the initial voltage of C2 to −1.506 V.

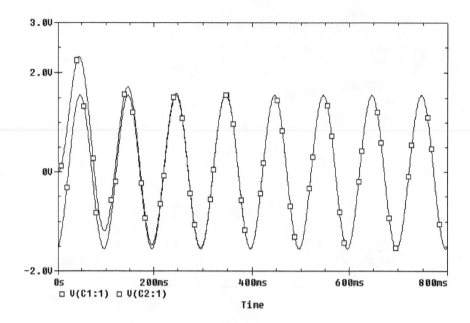

Figure 4.19 Plots of the voltages of the capacitors in Figure 4.17.

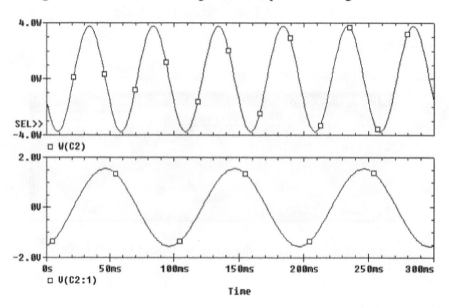

Figure 4.20 Plots of the capacitor power (top) and voltages (bottom).

Step 5 Verify that the simulation results are correct.

The capacitor voltage

   (1) is a sinusoidal function of time
   (2) has an amplitude of 1.55 V
   (3) has a period of 0.1 seconds

and

   (4) has an initial value of $v_c(0) = -1.506$ V.

The plot of the capacitor voltage in Figure 4.20 has all four of these features. The simulation results are correct.

Step 6 Report the answer to the circuit analysis problem.

Figure 4.20 exhibits the required plot of the power delivered to the capacitor as a function of time, t, and also the required plot of the capacitor voltage, $v_c(t)$, versus t.

4.5 SUMMARY

PSpice can be used to simulate circuits in the time domain. Time domain analysis is appropriate for circuits that contain capacitors and/or inductors and either are excited by time varying inputs or contain switches.

PSpice provides voltage and current sources to simulate time varying inputs as shown in Table 4.1. PSpice also provides parts to simulate switches as shown in Table 4.2.

CHAPTER 5

Analysis of AC Circuits

An AC circuit is a linear, time-invariant circuit that is excited by a sinusoidal input. The input to the AC circuit is generally the voltage of an independent voltage source or the current of an independent current source. An AC circuit can have more than one input, provided that all inputs are sinusoids having the same frequency. The output of the AC circuit can be the steady-state voltage or current of any element of the circuit. The output of an AC circuit is sinusoidal and has the same frequency as the input.

AC circuits can be analyzed using impedances and phasors. Phasors are complex numbers that represent the amplitude and phase angle of a sinusoidal current or voltage. The correspondence between a sinusoid and phasor is

$$A\cos(\omega t + \theta) \quad \leftrightarrow \quad Ae^{j\theta} = A\angle\theta$$

PSpice will calculate the phasor representing any current or voltage in an AC circuit.

5.1 AC Circuits

Table 5.1 shows the new parts needed to analyze AC circuits using PSpice. The first two rows of Table 5.1 show AC sources. These sources describe the phasors that represent sinusoidal sources. The amplitude and phase angle are properties of the source and are set using the property editor. The frequency is specified separately, using the "Simulation Settings" dialog box.

The last three rows of Table 5.1 show printers. These printers each cause a phasor to be printed in the PSpice output file. The printers are connected to a circuit in the same way that voltmeters and ammeters are connected to a circuit. For example, Figure 5.1 shows three printers connected to a circuit. The printer labeled IPRINT is a IPRINT part and so measures a current. In Figure 5.1, this printer measures the inductor current. The other two printers in Figure 5.1 are not labeled. They are VPRINT parts and so measure voltages. The printer with only one lead measures the node voltage at the top node of capacitor C1. The VPRINT part with two leads measures the voltage across resistor R2.

Notice the minus sign on the symbol for the printers. This minus sign marks the printer terminal corresponding to the minus sign of the voltage polarity. Consequently, the VPRINT printer connected to resistor R2 measures the voltage having the polarity: − on top, + on bottom. Printer currents and voltages adhere to the passive convention. Consequently, the IPRINT printer measures the current directed from node L1:1 to node L1:2.

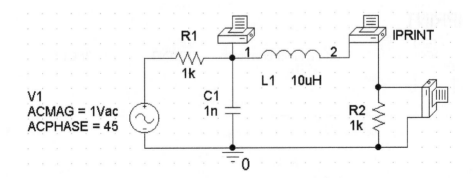

Figure 5.1 Printers are connected like voltmeters or ammeters.

Table 5.1 PSpice parts for AC circuits and the libraries in which they are found

Symbol	Description	PSpice Name	Library
1Vac 0Vdc V?	AC voltage source	VAC	SOURCE
1Aac 0Adc I?	AC current source	IAC	SOURCE
	Print element voltage	VPRINT2	SPECIAL
	Print node voltage	VPRINT1	SPECIAL
IPRINT	Print element current	IPRINT	SPECIAL

Example 5.1 Figure 5.2*a* shows an AC circuit. The input to this circuit is the voltage of the voltage source. The currents $i_1(t)$ and $i_2(t)$ and the voltage $v(t)$ have been identified as outputs. In Figure 5.2*b* the same circuit has been represented in the frequency domain using phasors and impedances. $\mathbf{I}_1(\omega)$, $\mathbf{I}_2(\omega)$ and $\mathbf{V}(\omega)$ are the phasors corresponding to $i_1(t)$, $i_2(t)$ and $v(t)$.

Step 1 Formulate a circuit analysis problem.

Analyze the circuit shown in Figure 5.2*a* to determine the currents $i_1(t)$ and $i_2(t)$ and the voltage $v(t)$.

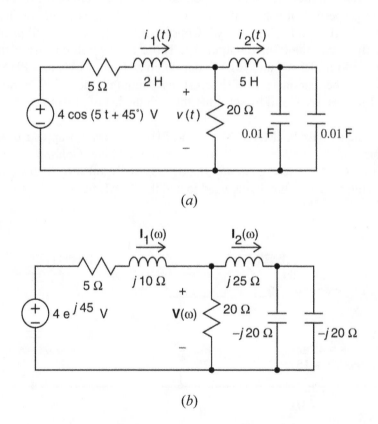

Figure 5.2 A circuit represented in the (*a*) time domain and (*b*) frequency domain.

Step 2 Describe the circuit using OrCAD Capture.

Figure 5.3 shows the circuit as described in Capture. It includes four parts from Table 5.1, the AC voltage source and the three printers.

The voltage source is a VAC part as shown in row 1 of Table 5.1. The magnitude and phase angle of the input voltage are specified by editing the VAC part properties using the property editor as shown in Figure 5.4. Also, the property editor has been used to change which properties of the VAC part are displayed. The ACMAG and ACPHASE properties of the VAC part are shown in Figure 5.3.

The properties of each printer must be edited to specify what is to be printed. Left-click on a printer to select it then right-click to pop up a menu. Select "Edit Properties." Figure 5.5 shows the property editor as it is used to set the set the contents of the fields AC, REAL, IMAG, MAG and PHASE to "y." Consequently, the printer will print the results of the AC analysis into the PSpice output file. Those results will consist of the phasor in both rectangular form, REAL and IMAG, and in polar form, MAG and PHASE. (Figure 5.5 shows only the parameters of interest by "hiding" the other parameters. The parameters of interest are found by scrolling through the list of parameters.)

(If one of the fields REAL, IMAG, MAG and PHASE does not appear in the property editor for some of the printers, add it by pressing the "New Column…" button in the property editor. A "Add New Column" dialog box will pop up. Figure 5.6 shows the "Add New Column" dialog box being used to add the REAL field.)

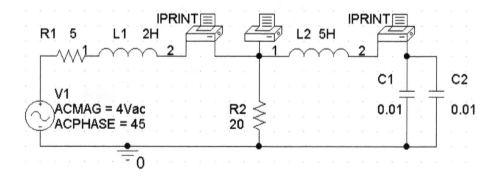

Figure 5.3 The circuit described in OrCAD Capture.

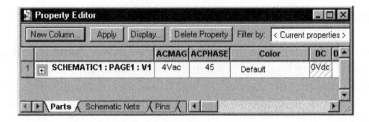

Figure 5.4 Editing the properties of the AC voltage source.

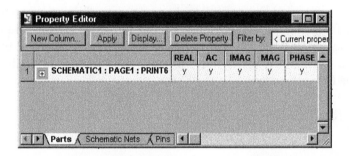

Figure 5.5 Editing printer properties.

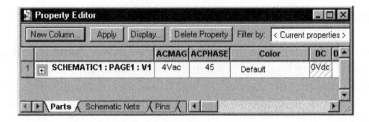

Figure 5.6 Adding a field in the property editor.

Step 3   Simulate the circuit using PSpice.

Select **PSpice / New Simulation Profile** from the OrCAD Capture menus to pop up the "New Simulation" dialog box. Specify a simulation name, then select "Create" to close "New Simulation" dialog box and pop up the "Simulation Settings" dialog box as shown in Figure 5.7. In the "Simulation Settings" dialog box, select "AC Sweep/Noise" as the analysis type.

AC Sweep analysis allows us to simulate a circuit over a range of frequencies. In this example, we only want to simulate the circuit at a single frequency so we will set the "Start Frequency" and "End Frequency" to the same value. PSpice uses units of Hertz for frequency. The frequency of the sinusoidal source in Figure 5.2a is 5 rad/s = 0.795775 Hertz. Set the values of both "Start Frequency" and "End Frequency" to 0.795775 as shown in Figure 5.7. Finally, select "OK" to close the "Simulation Settings" dialog box and return to the Capture screen.

Select **PSpice / Run** from the OrCAD Capture menus to run the simulation.

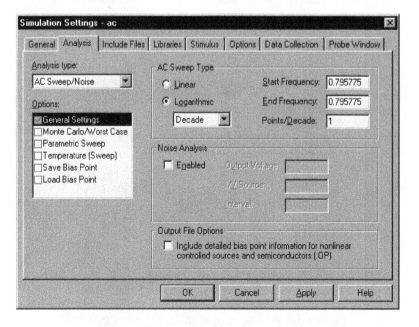

Figure 5.7 The "Simulation Settings" dialog box.

Step 4 **Display** the results of the simulation, for example, using Probe.

After a successful AC Sweep/Noise simulation, Probe, the graphical post-processor for PSpice, will open automatically in a Schematics window. Select **View / Output File** from the Schematics menus. Scroll through the output file to find results of the AC analysis shown in Figure 5.8. These results indicate that

$$\mathbf{I}_1(\omega) = 0.1733\angle{-13.1°} = 0.1688 - j\,0.03927,$$

$$\mathbf{V}(\omega) = 2.079\angle{40.03°} = 1.592 + j\,1.337$$

and

$$\mathbf{I}_2(\omega) = 0.1368\angle{-49.97°} = 0.08915 - j\,0.1061$$

(The names V_PRINT1, N00877 and V_PRINT3 that PSpice used to identify the printers are not very helpful. These names can be changed to something more intuitive by editing the "Reference" property of the printer.)

Step 5 **Verify** that the simulation results are correct.

Apply Kirchhoff's current law at the top node of the 20 Ω resistor in Figure 5.2*b* to get

$$\mathbf{I}_1(\omega) = \frac{\mathbf{V}(\omega)}{20} + \mathbf{I}_2(\omega)$$

```
****        AC ANALYSIS              TEMPERATURE =    27.000 DEG C


FREQ          IM(V_PRINT1)IP(V_PRINT1)IR(V_PRINT1)II(V_PRINT1)
7.958E-01     1.733E-01   -1.310E+01   1.688E-01   -3.927E-02


FREQ          VM(N00877)   VP(N00877)   VR(N00877)   VI(N00877)
7.958E-01     2.079E+00    4.003E+01    1.592E+00    1.337E+00


FREQ          IM(V_PRINT3)IP(V_PRINT3)IR(V_PRINT3)II(V_PRINT3)
7.958E-01     1.386E-01   -4.997E+01   8.915E-02   -1.061E-01
```

Figure 5.8 AC analysis results from the PSpice output file.

Substituting the simulation results shows

$$0.1688 - j0.03927 = \mathbf{I}_1(\omega) = \frac{\mathbf{V}(\omega)}{20} + \mathbf{I}_2(\omega)$$

$$= \frac{1.592 + j1.337}{20} + 0.08915 - j0.1061$$

$$= 0.1672 - j0.03925$$

The error in the real part is less than 1% and the error in the imaginary part is much smaller than that. The simulation results satisfy this Kirchhoff's current law equation.

It's likely that the simulation results are correct. To be sure, apply Kirchhoff's voltage law to the left-hand mesh to get

$$4e^{j45} = (5 + j10)\mathbf{I}_1(\omega) + \mathbf{V}(\omega)$$

Substituting the simulation results shows

$$2.8284 + j2.8284 = (5 + j10)(0.1688 - j0.03927) + (1.592 + j1.337)$$

$$= 1.2367 + j1.4917 + (1.592 + j1.337)$$

$$= 2.8287 + j2.8287$$

The simulation results satisfy this Kirchhoff's voltage law equation. They are correct.

Step 6 Report the answer to the circuit analysis problem.

The outputs of the circuit shown in Figure 5.2a are

$$i_1(t) = 0.1733 \cos(5t - 13.1°)\ \text{A},$$

$$v(t) = 2.079 \cos(5t + 40.03°)\ \text{V}$$

and

$$i_2(t) = 0.1368 \cos(5t - 49.97°)\ \text{A}$$

5.2 Coupled Coils

Table 5.2 shows PSpice parts used to represent magnetically coupled coils. Magnetically coupled coils exhibit both self-inductance and mutual inductance. In PSpice, the mutual inductance of coupled coils is specified using the coupling coefficient k. For example, suppose two coils having self-inductances L_1 and L_2 are coupled by a mutual inductance M. The coupling coefficient, k, is given by

$$0 \le k = \frac{M}{\sqrt{L_1 L_2}} \le 1 \tag{5.1}$$

Three steps are required to represent coupled coils. First add inductors, the parts representing the coils, to the circuit. The self-inductances of the coupled coils are the inductances of the inductors representing the coils. Edit the properties of the inductors to specify the self-inductances of the coupled coils. Second, place a K_Linear part next to the circuit. The K_Linear part is shown in the first row of Table 5.2. (The K_Linear part is not wired to the rest of the circuit.) Third, edit the properties of the K_Linear part to specify the value of the coupling coefficient and to indicate which coils are coupled.

Table 5.2 PSpice parts for coupled coils and transformers

Symbol	Description	PSpice Name	Library
K K? K_Linear COUPLING = 1	Mutual Inductance	K_Linear	ANALOG
TX?	Transformer	XFRM_ LINEAR	ANALOG

Example 5.2 Figure 5.9*a* shows a circuit that includes coupled coils. Designate the horizontal inductor to be L_1 and the vertical inductor to be L_2. Then $L_1 = 4$ H and $L_2 = 5$ H. The mutual inductance is M = 3 H. From Equation 5.1, the coupling coefficient, k is given by

$$0 \leq k = \frac{M}{\sqrt{L_1 L_2}} = \frac{3}{\sqrt{4 \times 5}} = 0.671 \leq 1$$

In Figure 5.9*b* the same circuit has been represented in the frequency domain using phasors and impedances. $I_1(\omega)$, $I_2(\omega)$, $V_1(\omega)$ and $V_2(\omega)$ are the phasors corresponding to $i_1(t)$, $i_2(t)$, $v_1(t)$ and $v_2(t)$.

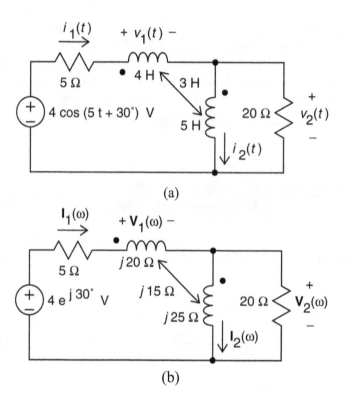

(a)

(b)

Figure 5.9 A circuit containing coupled coils. The circuit is represented in the (*a*) time domain and (*b*) frequency domain.

109

Step 1 Formulate a circuit analysis problem.

Analyze the circuit shown in Figure 5.9*a* to determine the currents $i_1(t)$ and $i_2(t)$ and the voltages $v_1(t)$ and $v_2(t)$.

Step 2 Describe the circuit using OrCAD Capture.

Figure 5.10 shows the circuit as described in Capture. The K_Linear part representing the mutual inductance has been placed below the circuit and is not connected to the other parts by wires.

In Figure 5.11 the property editor is used to specify the value of the coupling coefficient and to indicate which coils are coupled. (Invoke the property editor by left-clicking on the K_Linear part to select it, then right-clicking to pop up a menu. Select "Edit Properties.")

Figure 5.10 shows that printers have been added to print the values of $I_1(\omega)$, $I_2(\omega)$, $V_1(\omega)$ and $V_2(\omega)$ into the PSpice output file. ($I_1(\omega)$, $I_2(\omega)$, $V_1(\omega)$ and $V_2(\omega)$ are the phasors corresponding to $i_1(t)$, $i_2(t)$, $v_1(t)$ and $v_2(t)$.)

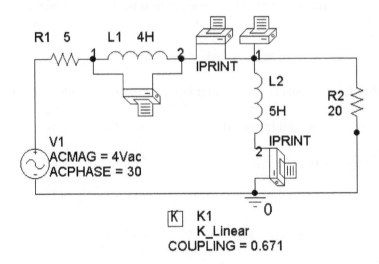

Figure 5.10 The circuit described in OrCAD Capture.

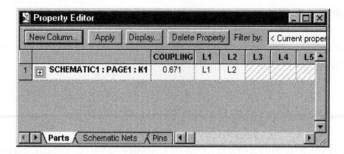

Figure 5.11 Specifying the coupling coefficient using the property editor.

Step 3 Simulate the circuit using PSpice.

Select **PSpice / New Simulation Profile** from the OrCAD Capture menus to pop up the "New Simulation" dialog box. Specify a simulation name, then select "Create" to close the "New Simulation" dialog box and pop up the "Simulation Settings" dialog box. In the "Simulation Settings" dialog box, select "AC Sweep/Noise" as the analysis type. We only want to simulate the circuit at a single frequency so we will set the "Start Frequency" and "End Frequency" to the same value 5 rad/s = 0.795775 Hertz. Finally, select "OK" to close the "Simulation Settings" dialog box and return to the Capture screen.

Select **PSpice / Run** from the OrCAD Capture menus to run the simulation.

Step 4 Display the results of the simulation, for example, using Probe.

After a successful AC Sweep/Noise simulation, Probe will open automatically in a Schematics window. Select **View / Output File** from the Schematics menus. Scroll through the output file to find results of the AC analysis. These results indicate that

$$\mathbf{I}_1(\omega) = 0.07835\angle{-14.8°} = 0.07575 - j\,0.02002,$$

$$\mathbf{I}_2(\omega) = 0.06119\angle{-103.0°} = -0.01378 - j\,0.05961,$$

$$\mathbf{V}_1(\omega) = 1.841\angle{45.3°} = 1.295 + j\,1.308$$

and

$$\mathbf{V}_2(\omega) = 1.958\angle{23.86°} = 1.791 + j\,0.7919$$

Step 5 Verify that the simulation results are correct.

The coupled coils in Figure 5.8*b* are represented by the equations

$$\mathbf{V}_1(\omega) = j\,20\,\mathbf{I}_1(\omega) + j\,15\,\mathbf{I}_2(\omega)$$
$$\mathbf{V}_2(\omega) = j\,25\,\mathbf{I}_2(\omega) + j\,15\,\mathbf{I}_1(\omega)$$

Substituting the values of the currents $\mathbf{I}_1(\omega)$ and $\mathbf{I}_2(\omega)$ obtained using PSpice gives

$$
\begin{aligned}
\mathbf{V}_1(\omega) &= j\,20\,(0.07575 - j0.02002) + j\,15\,(-0.01378 - j0.05961) \\
&= (\,j1.515 + 0.4004) + (-j0.2067 + 0.89415) \\
&= j1.308 + 1.29455 \\
\mathbf{V}_2(\omega) &= j\,25\,(-0.01378 - j0.05961) + j\,15\,(0.07575 - j0.02002) \\
&= (-j0.3445 + 1.49025) + (\,j1.13625 + 0.3003) \\
&= j0.79175 + 1.79055
\end{aligned}
$$

These values agree with the values of $\mathbf{V}_1(\omega)$ and $\mathbf{V}_2(\omega)$ obtained using PSpice. The simulation results are correct.

Step 6 Report the answer to the circuit analysis problem.

The currents and voltages of the coupled coils in the circuit shown in Figure 5.8*a* are

$$i_1(t) = 0.07835 \cos(5t - 14.8°)\ \text{A},$$

$$i_2(t) = 0.06119 \cos(5t - 103.0°)\ \text{A}$$

$$v_1(t) = 1.841 \cos(5t + 45.3°)\ \text{V}$$

and

$$v_2(t) = 1.958 \cos(5t + 23.86°)\ \text{V}$$

5.3 Ideal Transformers

The XFRM_LINEAR part shown in the second row of Table 5.2 can also be used to represent coupled coils. The properties of the XFRM_LINEAR include the inductance of each coil, L_1 and L_2, and the coupling coefficient, k. When $k = 1$ and, also, L_1 and L_2 are large enough that the impedances of the coils are much larger than the other impedances in the circuit, the XFRM_LINEAR will represent an ideal transformer. The turns ratio of the ideal transformer is related to the coil inductances by

$$n = \frac{N_2}{N_1} = \sqrt{\frac{L_2}{L_1}} \tag{5.2}$$

where N_1 is the number of turns in the coil having inductance L_1 and N_2 is the number of turns in the coil having inductance L_2.

Example 5.3 Figure 5.12a shows a circuit that includes an ideal transformer. Ideal transformers can be represented using the PSpice part called XFRM_LINEAR. XFRM_LINEAR is shown in the second row of Table 5.2.

Step 1 Formulate a circuit analysis problem.

Analyze the circuit shown in Figure 5.12a to determine the currents $i_1(t)$ and $i_2(t)$ and the voltages $v_1(t)$ and $v_2(t)$.

Step 2 Describe the circuit using OrCAD Capture.

Figure 5.13 shows the circuit described in Capture. The device TX1 is a XFRM_LINEAR part representing the ideal transformer. We need to specify the coil inductances and the coupling coefficient of TX1. The value of the coupling coefficient of an ideal transformer is $k = 1$. Figure 5.12a indicates that the turns ratio of the ideal transformer is $n = 2$. Equation 5.2 requires that $L_2 = 4 L_1$ to obtain a turns ratio equal to 2. Figure 5.12b indicates that the largest impedance in the circuit is 5 Ω. Choosing $L_1 = 10$ H and $L_2 = 40$ H cause the coil impedances to be $j100$ Ω and $j400$ Ω, both much larger than 5 Ω.

In Figure 5.14 the property editor is used to specify the value of the coupling coefficient and the values of the coil inductances.

The circuit shown in Figure 5.12a consists of two separate parts. One part is comprised of the voltage source, capacitor, the 5 Ω resistor and the primary (left) coil of the transformer. The other part is comprised of the 20 mH inductor, the 2 Ω resistor and the secondary (right) coil of the transformer. These two parts are connected magnetically, but not electrically. No charge flows from one part to the other. PSpice will only analyze circuits that consist of a single part, electrically. The resistor R3 is added in Figure 5.13 to connect the two parts electrically. The resistance of R3 is very large so that R3 acts like an open circuit and the circuit in Figure 5.13 will behave like the circuit in Figure 5.12a.

Figure 5.13 shows that printers have been added to print the values of $I_1(\omega)$, $I_2(\omega)$, $V_1(\omega)$ and $V_2(\omega)$ into the PSpice output file.

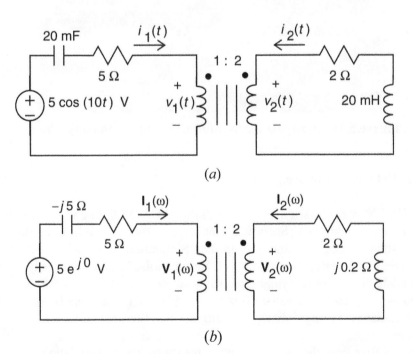

(a)

(b)

Figure 5.12 A circuit represented in the (a) time and (b) frequency domain.

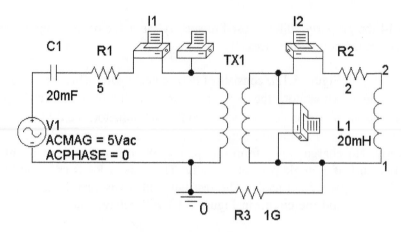

Figure 5.13 The circuit described in OrCAD Capture.

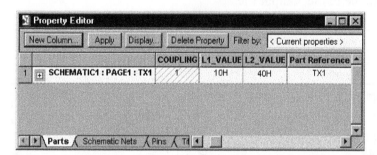

Figure 5.14 Specifying the transformer using the property editor.

Step 3 Simulate the circuit using PSpice.

Select **PSpice / New Simulation Profile** from the OrCAD Capture menus to pop up the "New Simulation" dialog box. Specify a simulation name, then select "Create" to close "New Simulation" dialog box and pop up the "Simulation Settings" dialog box. In the "Simulation Settings" dialog box, select "AC Sweep/Noise" as the analysis type. We only want to simulate the circuit at a single frequency so we will set the "Start Frequency" and "End Frequency" to the same value 10 rad/s = 1.591549 Hertz. Finally, select "OK" to close the "Simulation Settings" dialog box and return to the Capture screen.

Select **PSpice / Run** from the OrCAD Capture menus to run the simulation.

Step 4 **Display** the results of the simulation, for example, using Probe.

After a successful AC Sweep/Noise simulation, Probe, the graphical post-processor for PSpice, will open automatically in a Schematics window. Select **View / Output File** from the Schematics menus. Scroll through the output file to find results of the AC analysis. These results indicate that

$$\mathbf{I}_1(\omega) = 0.6759\angle 41.98° = 0.5.025 - j\,0.45.21,$$

$$\mathbf{I}_2(\omega) = 0.3378\angle -137.7° = -0.2500 - j\,0.2272$$

$$\mathbf{V}_1(\omega) = 0.3395\angle 47.97° = 0.2273 + j\,0.2522,$$

and

$$\mathbf{V}_2(\omega) = 0.6789\angle 47.97° = 0.4545 + j\,0.5043$$

Step 5 **Verify** that the simulation results are correct.

The ideal transformer in Figure 5.12*b* is represented by the equations

$$\frac{\mathbf{V}_2(\omega)}{\mathbf{V}_1(\omega)} = n = 2 \quad \text{and} \quad \frac{\mathbf{I}_2(\omega)}{\mathbf{I}_1(\omega)} = -\frac{1}{n} = -0.5$$

Substituting the values of the voltages $\mathbf{V}_1(\omega)$ and $\mathbf{V}_2(\omega)$ obtained using PSpice gives

$$\frac{0.6789\angle 47.97°}{0.3395\angle 47.97°} = 1.9997\angle 0 \approx 2$$

and substituting the values of the currents $\mathbf{I}_1(\omega)$ and $\mathbf{I}_2(\omega)$ obtained using PSpice gives

$$\frac{0.3378\angle -137.7°}{0.6759\angle 41.98°} = 0.4998\angle -179.68° \approx -0.5$$

The simulation results are correct.

Step 6 Report the answer to the circuit analysis problem.

The currents and voltages of the ideal transformer in the circuit shown in Figure 5.12a are

$$i_1(t) = 0.6759 \cos (5t + 41.98°) \text{ A},$$

$$i_2(t) = 0.3378 \cos (5t - 137.7°) \text{ A}$$

$$v_1(t) = 0.3395 \cos (5t + 47.97°) \text{ V}$$

and

$$v_2(t) = 0.6789 \cos (5t + 47.97°) \text{ V}$$

5.4 SUMMARY

PSpice can be used to analyze an AC circuit. The phasors corresponding to the currents and voltages of AC circuits are printed into the PSpice output file using the printers shown in Table 5.1.

Coupled coils and transformers can be incorporated into AC circuits using the PSpice part K_Linear and XFRM_LINEAR shown in Table 5.2.

CHAPTER 6

Frequency Response

Suppose $v_s(t) = A\cos(\omega t + \theta)$ V is the input to a linear, time-invariant circuit and $v_o(t) = B\cos(\omega t + \phi)$ V is the steady-state response of the circuit. The network function, $\mathbf{H}(\omega)$, is the ratio of the response phasor to the input phasor

$$\mathbf{H}(\omega) = \frac{\mathbf{V}_o(\omega)}{\mathbf{V}_s(\omega)} = \frac{A\angle\theta}{B\angle\phi} = \frac{A}{B}\angle(\theta - \phi)$$

The magnitude of the network function, $\left|\mathbf{H}(\omega)\right| = \dfrac{A}{B}$ is called the gain of the circuit and the angle of the network $\angle\mathbf{H}(\omega) = \theta - \phi$ is called the phase shift of the circuit. The gain and phase shift are both functions the frequency ω. These two functions constitute the frequency response of the circuit. PSpice represents the frequency response graphically by providing graphs of the gain versus frequency and of the phase shift versus frequency. These graphs are called frequency response plots.

6.1 Frequency Response Plots

The frequency axis of a frequency response plot can be either a linear axis or a logarithmic axis. When a logarithmic axis is used for the frequency variable, the plots are referred to as Bode diagrams or Bode plots.

We encounter the terms "octave" and "decade" when working with logarithmic scales. The frequency doubles in an octave and increases by a factor of ten in a decade. (The log of the frequency increases by 1 as the frequency increases by a decade.)

Let $A\angle\theta$ be the phasor of the node voltage at node 2 of a circuit. PSpice uses the notation:

$$V(2)\angle Vp(2) = A\angle\theta$$

That is, V(2) denotes the magnitude of the phasor and Vp(2) denotes the angle of the phasor. PSpice gives the angle in degrees. Similarly, V(R2) represents the magnitude of the voltage across resistor R2 while Vp(R2) denotes the angle.

In Bode plots, the gain is often expressed in units of decibels (dB), defined as

$$\left| \mathbf{H}(\omega) \right|_{dB} = 20 \log_{10} \left| \mathbf{H}(\omega) \right|$$

PSpice indicates that the units are decibels by inserting "dB" into the name of a signal just before the parenthesis. For example, VdB(2) denotes the magnitude of the node voltage phasor, in dB.

Example 6.1 The input to the circuit shown in Figure 6.1 is the voltage source voltage $v_s(t)$. The response is the voltage, $v_o(t)$, across the 20 kΩ resistor. The network function of this circuit is

$$\mathbf{H}(\omega) = \frac{\mathbf{V}_o(\omega)}{\mathbf{V}_s(\omega)} = -\frac{R_2}{R_1 + j\omega C R_1 R_2} = \frac{-4}{1 + j\,0.008\,\omega}$$

Graph the frequency response of this circuit.

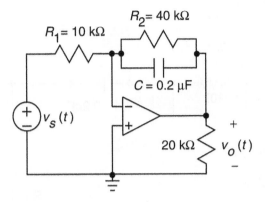

Figure 6.1 A circuit.

Step 1 Formulate a circuit analysis problem.

Obtain the gain and phase Bode plots of the circuit shown in Figure 6.1.

Step 2 Describe the circuit using OrCAD Capture.

Figure 6.2 shows the circuit as described in Capture.

Two nodes of this circuit have been named using the using a PSpice part called an "off-page connector." The particular off-page connector used in Figure 6.2 is called a "OFFPAGELEFT-R" part and is found in the part library named CAPSYM. To label a node, select **Place / Off-Page Connector…** from the OrCAD capture menus to pop up the "Place Off-Page Connector" dialog box as shown in Figure 6.3. Select the library CAPSYM from the list of libraries. (If CAPSYM does not appear in the list of libraries, left-click the "Add Libraries" button to pop up a "Browse File" dialog box. Select capsym.olb from the list of libraries in the "Libraries" folder, then left-click the "Open" button to return to "Place Off-Page Connector" dialog box.) Left-click the "OK" button to place the part in the OrCAD workspace.

The new connector will be labeled as "OFFPAGELEFT-R." Use the property editor to change this name to something descriptive such as "Vo." Wire the connector to the appropriate node of the circuit to name that node "Vo."

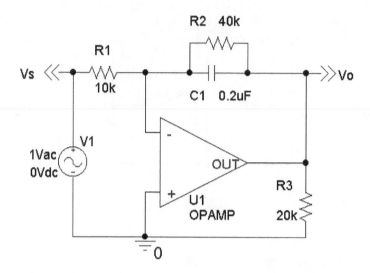

Figure 6.2 The circuit as described in OrCAD Capture.

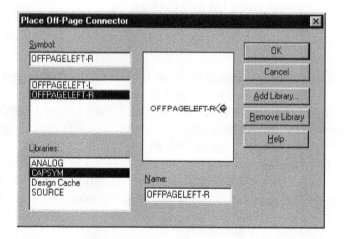

Figure 6.3 The Place Off-Page Connector dialog box.

Step 3 **Simulate** the circuit using PSpice.

Select **PSpice / New Simulation Profile** from the OrCAD Capture menus to pop up the "New Simulation" dialog box. Specify a simulation name, then select "Create" to close the "New Simulation" dialog box and pop up the "Simulation Settings" dialog box as shown in Figure 6.4. In the "Simulation Settings" dialog box, select "AC Sweep/Noise" as the analysis type.

AC Sweep analysis allows us to simulate a circuit over a range of frequencies. Some trial and error may be required to determine an appropriate range of frequencies. In this example, the network function has only one corner frequency, a pole at $\frac{1}{0.008} = 125$ rad/s $= 19.89$ Hertz. It's reasonable to use a "Start Frequency" of 1 Hertz and an "End Frequency" of 1000 Hertz. Select a logarithmic frequency axis to graph the Bode plots. Finally, select "OK" to close the "Simulation Settings" dialog box and return to the Capture screen.

Select **PSpice / Run** from the OrCAD Capture menus to run the simulation.

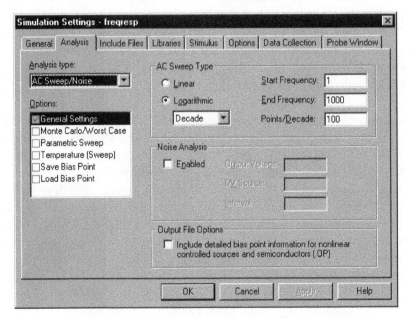

Figure 6.4 The Simulation Settings dialog box.

Step 4 Display the results of the simulation, for example, using Probe.

After a successful AC Sweep simulation, Probe, the graphical post-processor, will open automatically in a Schematics window. Select **Plot / Add Plot to Window** from the Schematics menus to add a second plot. Two empty plots will appear, one above the other. Select the top plot by left-clicking the mouse on the top plot.

Select **Trace / Add Trace** from the Schematics menus to pop up the "Add Traces" dialog box shown in Figure 6.5. Select first V(Vo) and then V(Vs) from the list of "Simulation Output Variables." The "Trace Expression," near the bottom of the dialog box, will be "V(Vo)V(Vs)." Edit the trace expression to be "Vdb(Vo)-Vdb(Vs)." Because

$$20 \log_{10} \frac{\mathbf{V}_o(\omega)}{\mathbf{V}_s(\omega)} = 20 \log_{10} \mathbf{V}_o(\omega) - 20 \log_{10} \mathbf{V}_s(\omega)$$

Vdb(Vo)-Vdb(Vs) is the gain in decibels. Left-click "OK" to close the "Add Traces" dialog box.

Select the bottom plot by left-clicking the mouse on the bottom plot. Select **Trace / Add Trace** to pop up the "Add Traces" dialog box. Select first V(Vo) and then V(Vs) from the list of "Simulation Output Variables." The "Trace Expression," near the bottom of the dialog box, will be "V(Vo)V(Vs)." Edit the trace expression to be "Vp(Vo)–Vp(Vs)." Vp(Vo)–Vp(Vs) is the phase shift in degrees. Left-click "OK" to close the "Add Traces" dialog box. The Bode plots are shown in Figure 6.6.

Probe provides tools to label some points on the Bode plots. Select the top plot, the gain Bode plot, by left-clicking on it. Select **Trace / Cursor / Display** from the Schematic menus to activate the cursors. (This is a toggle, so selecting **Trace / Cursor / Display** will also deactivate an active cursor.) The cursor is positioned using the arrow keys or by left-clicking on a trace using the mouse. Move the cursor to a low frequency (for example, 1.1235 Hertz is used in Figure 6.5) to measure the low frequency gain of the circuit. When the cursor is positioned as desired, select **Plot / Label / Mark** from the Schematics menus to mark the point. Deactivate the cursor by selecting **Trace / Cursor / Display** from the Schematics menus. Then position the label as desired using the mouse. The point on the gain Bode plot is labeled as

(1.1235, 12.027)

This label indicates that at a frequency of 1.1235 Hertz the gain is 12.027 dB. The gain will be 3 dB lower at the corner frequency. Activate the cursor and use it to locate and label the point on the gain Bode plot where the gain is 9.027 dB. (The gain changes from one value to another in small increments as the cursor moves along the trace. Figure 6.6 shows that the gain is 9.0296 dB, the available value closest to 9.041 dB, at a frequency of 19.850 Hertz.)

Select the bottom plot, the phase Bode plot, by left-clicking on it. Activate the cursor and use it to locate and label the point on the phase Bode plot where the angle is 135°. The point on the phase Bode plot is labeled as (19.850, 135.065). This label indicates that at a frequency of 19.850 Hertz the phase shift is 135.065°. The labeled plot is shown in Figure 6.6.

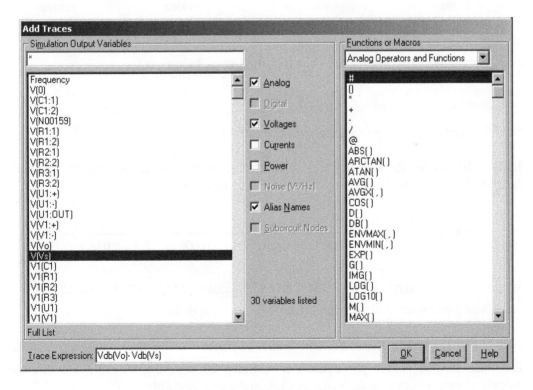

Figure 6.5 The Add Traces dialog box.

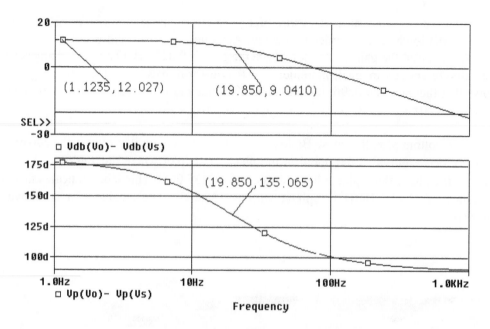

Figure 6.6 The Gain and Phase Bode plots.

Step 5 Verify that the simulation results are correct.

Let's check to see that the points labeled on the Bode plots agree with the network function of the circuit. The network function of the circuit is

$$\mathbf{H}(\omega)=\frac{\mathbf{V}_o(\omega)}{\mathbf{V}_s(\omega)}=-\frac{R_2}{R_1+j\omega C R_1 R_2}=\frac{-4}{1+j0.008\,\omega}=\frac{-4}{1+j0.016\pi f}$$

When f = 1.1235 Hertz, $\mathbf{H}(\omega)=3.994\angle176.7°$. Since 20 $\log_{10}$ 3.994 = 12.027 dB, the calculated gain agrees with the value labeled on the gain Bode plot.

Also, when f = 19.850 Hertz, $\mathbf{H}(\omega)=2.8316\angle135.064°$. The phase angle agrees with the value labeled on the phase Bode plot. Since 20 $\log_{10}$ 2.8316 = 9.0406 dB, the calculated gain agrees with the value labeled on the gain Bode plot.

The Bode plots agree with the network function so the simulation results are correct.

Step 6 Report the answer to the circuit analysis problem.

Figure 6.6 shows the Bode plots of the circuit shown in Figure 6.1.

Example 6.2 This example illustrates that global parameters can be used to incorporate design equations into a PSpice circuit. It also illustrates that the LAPLACE part provided by PSpice makes it easy to verify that a frequency response corresponds to a specified transfer function.

The circuit shown in Figure 6.7 is a standard filter circuit called a Sallen-Key lowpass filter. The transfer function of this circuit is

$$T(s) = \frac{V_o(s)}{V_i(s)} = \frac{k\omega_0^2}{s^2 + \dfrac{\omega_0}{Q}s + \omega_0^2}$$

where k is the dc gain, ω_0 is the corner frequency and Q is the quality factor of the filter. The Sallen-Key lowpass filter can be designed to have specified values of ω_0 and Q using the following equations.

$$C_1 = C_2 = C = 0.1\ \mu F \tag{6.1}$$

$$R_1 = R_2 = R_4 = R = \frac{1}{C\omega_0} \tag{6.2}$$

$$k = 3 - \frac{1}{Q} \tag{6.3}$$

$$R_3 = (k-1)R \tag{6.4}$$

In this example, these equations will be used to design a Sallen-Key lowpass filter to have $Q = 4$ and $\omega_0 = 1000*2\pi$ rad/s or, equivalently, $f_0 = 1000$ Hertz.

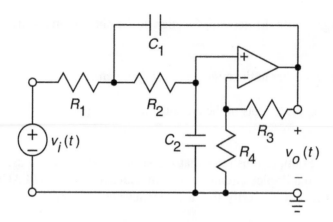

Figure 6.7 The Sallen-Key lowpass filter.

Step 1 Formulate a circuit analysis problem.

Design a Sallen-Key lowpass filter to have $Q = 4$ and $\omega_0 = 1000*2\pi$ rad/s. Obtain the gain Bode plot of the filter.

Step 2 Describe the circuit using OrCAD Capture.

Figure 6.8 shows the circuit as described in Capture.

Two things must be done in order to incorporate the design equations into the circuit. First, a PARAM part is placed in the drawing and its properties are edited to add the parameters C, Q, w0, R, k and Rk:

```
PARAMETERS:
C = 0.1uF
Q = 4
w0 = {1000*2*3.14159}
R = {1/(w0*C)}
k = {3-1/Q}
Rk = {(k-1)*R}
```

Second, the properties of the resistors and capacitors are edited to make their resistance and capacitance value dependent on these parameters.

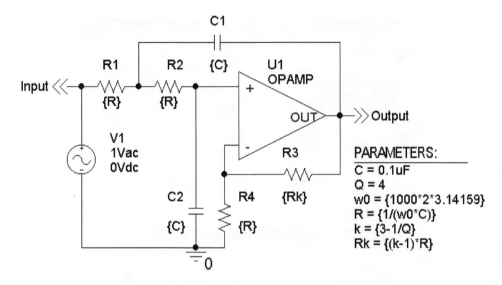

Figure 6.8 The Sallen-Key lowpass filter circuit as described in Capture.

To add the PARAM part to the drawing, select **Place / Parts** from the OrCAD Capture menus to open the "Place Parts" dialog box. The PARAM part is contained in the "SPECIAL" library. (It may be necessary to add this library. The file is called special.olb and resides in the "PSpice" folder.) Place a PARAM part in the OrCAD Capture workspace as shown in Figure 6.8.

Next select the PARAM part and edit its properties. (Left-click on the PARAM part, then right-click to pop up a menu. Select "Edit Properties…" from that menu.) A property sheet will open as shown in Figure 6.9. Left-click on the "New Column" button. A New Column dialog box will appear as shown in Figure 6.10. Supply the Name "w0" and the Value {1000*2*3.14159}. The braces are important. They indicate that the Value is a formula that is to be evaluated to determine the numerical value of w0. Click "OK" to close the "New Column" dialog box. Repeat this process to add all of the properties C, k, Q, R, Rk and w0 as shown in Figure 6.11. The order in which the properties are added is not important.

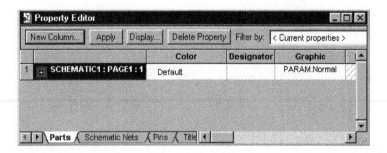

Figure 6.9 The Property Editor.

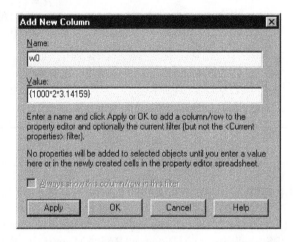

Figure 6.10 Adding the parameter w0.

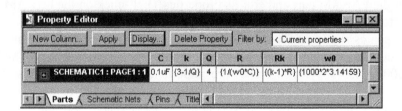

Figure 6.11 Adding the parameters.

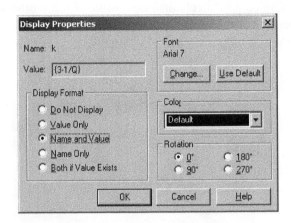

Figure 6.12 Displaying the parameters.

Notice that the Values of C and Q are numerical values while the values of k, R, Rk and w0 are formulas. The formulas for k, R and Rk involve the Names of other parameters. So, for example, the value of R depends on the values of w0 and C.

To display the Name and Value of a parameter on the circuit diagram, left-click on the name of the parameter in the property editor. For example, left-click on the Name "k" above the Value {3-1/Q} in Figure 6.11. Next, left-click the "Display..." button to pop up the "Display Properties" dialog box as shown in Figure 6.12. Select "Name and Value," then left-click the "OK" button to close the "Display Properties" dialog box. Repeat to display the names and values of the other parameters. Close the Property Editor.

Finally, use the property editor to specify values for the resistances and capacitances as functions of the parameters. For example, left-click on the value of capacitor C1 to pop up a "Display Properties" dialog box. Edit the Value to be {C} as shown in Figure 6.13. The braces are important. Close the "Display Properties" dialog box, then close the property editor.

Now the circuit drawing shown in Figure 6.8 is complete. The values of the parameters k, R, and Rk are calculated from the specified values of the parameters Q, C, and w0. The values of the resistances and capacitances R1, R2, R3, R4, C1 and C2 are calculated from the values of the parameters. Suppose the value of the parameter w0 was changed using the property editor. PSpice would calculate new values for the other parameters and then calculate new values for the circuit resistances and capacitances to cause the corner frequency of the circuit change to the new value of w0.

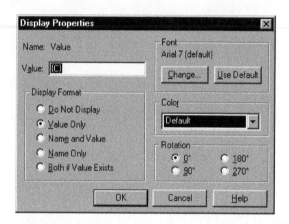

Figure 6.13 Capacitor C1 has a capacitance value equal to C.

Step 3 **Simulate** the circuit using PSpice.

Select **PSpice / New Simulation Profile** from the OrCAD Capture menus to pop up the "New Simulation" dialog box. Specify a simulation name, then select "Create" to close the "New Simulation" dialog box and pop up the "Simulation Settings" dialog box. In the "Simulation Settings" dialog box, select "AC Sweep/Noise" as the analysis type.

AC Sweep analysis allows us to simulate a circuit over a range of frequencies. In this example, the network function has only one corner frequency, at 1000 Hertz. It's reasonable to use a "Start Frequency" of 100 Hertz and an "End Frequency" of 10 kHertz. Select a logarithmic frequency axis to graph the Bode plots. Finally, select "OK" to close the "Simulation Settings" dialog box and return to the Capture screen.

Select **PSpice / Run** from the OrCAD Capture menus to run the simulation.

Step 4 **Display** the results of the simulation, for example, using Probe.

After a successful AC Sweep simulation, Probe, the graphical post-processor, will open automatically in a Schematics window. Select **Trace / Add Trace** to pop up the "Add Traces" dialog box shown in Figure 6.14 Select DB() from the "Functions or Macros" column then V(Output) from the "Simulation Output Variables" column. The Trace Expression will be DB(V(Output)), indicating the voltage at the output node in dB. Next, select DB() from the "Functions or Macros" column then V(Input) from the "Simulation Output Variables" column. Edit the trace expression to be "DB(V(Output))–DB(V(Input))," the gain in decibels. Left-click "OK" to close the "Add Traces" dialog box.

The gain Bode plot is shown in Figure 6.15.

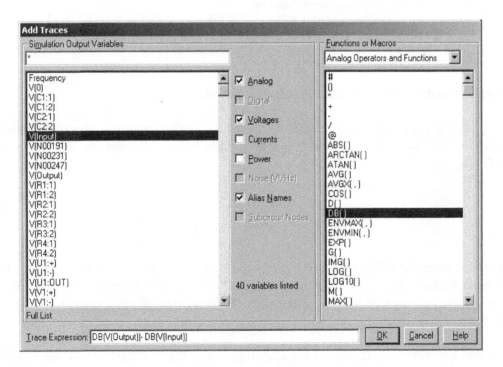

Figure 6.14 The Add Traces dialog box.

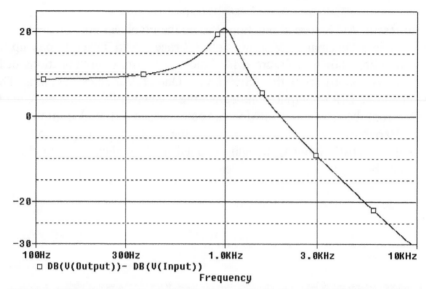

Figure 6.15 The magnitude Bode plot of the Sallen-Key lowpass filter.

Step 5 Verify that the simulation results are correct.

The PSpice part named LAPLACE simulates a circuit that has a specified transfer function. We can use the LAPLACE part to simulate a circuit that has the desired transfer function

$$T(s) = \frac{k\omega_0^{\,2}}{s^2 + \dfrac{\omega_0}{Q}s + \omega_0^{\,2}}$$

The gain bode plot of the LAPLACE part is the Bode plot that corresponds to the desired transfer function. If the gain bode plot of the LAPLACE part is identical to the gain Bode plot of the Sallen-Key lowpass filter, then the Bode plot of the Sallen-Key filter is correct.

To add the LAPLACE part to the drawing, select **Place / Parts** from the OrCAD Capture menus to open the "Place Parts" dialog box as shown in Figure 6.16. The

LAPLACE part is contained in the "ABM" library. (It may be necessary to add this library. The file is called abm.olb and resides in the "PSpice" folder.) Place a LAPLACE part in the OrCAD Capture workspace as shown in Figure 6.17.

The NUM and DENOM properties of the LAPLACE part represent the numerator and denominator polynomials of the transfer function. Both must be edited so that the LAPLACE part represents the desired transfer function. For example, to edit DENOM, left-click on the s+1 in the LAPLACE part, then right-click to pop up a menu. Select "Edit Properties…" from that menu to pop up a "Display Properties" dialog box. Edit the value of DENOM to be s*s+(w0/Q)+w0*w0 as shown in Figure 6.18. Change the display format to "Name and Value" also as shown in Figure 6.18. Close the "Display Properties" dialog box.

Figure 6.19 shows the circuit after the NUM and DENOM properties of the LAPLACE part have been edited. Figure 6.20 shows both the desired gain Bode plot and the gain Bode plot of the Sallen-Key filter. The plots are identical, overlapping exactly. The simulation results are correct.

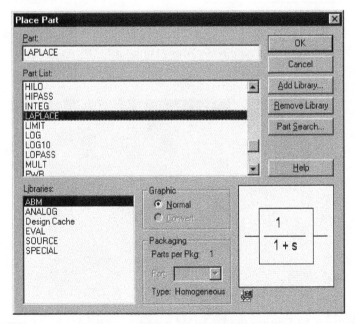

Figure 6.16 The LAPLACE part.

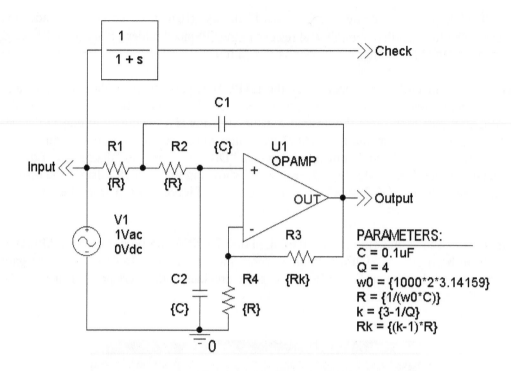

Figure 6.17 The circuit after placing a LAPLACE part.

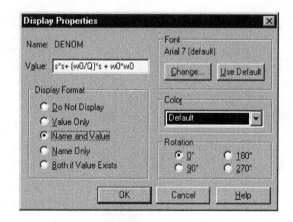

Figure 6.18 Editing the DENOM property of the LAPLACE part.

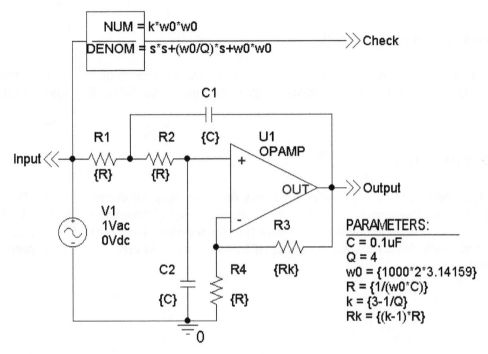

Figure 6.19 The circuit after editing the LAPLACE part properties.

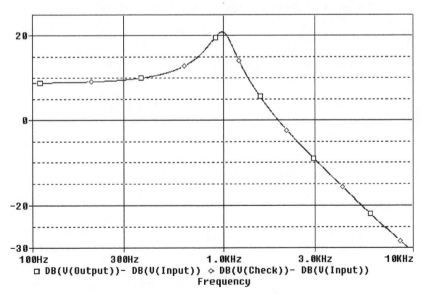

Figure 6.20 Comparing Bode plots.

Step 6 Report the answer to the circuit analysis problem.

The Sallen-Key lowpass filter shown in Figure 6.8 has a corner frequency at 1000 Hertz and a Q equal to 4. Figure 6.15 shows the gain Bode plot of the Sallen-Key lowpass filter.

6.2 SUMMARY

The AC Sweep analysis provides frequency response plots for electric circuits. Design equations are incorporated into PSpice circuits using global parameters. The LAPLACE part simulates a circuit having a specified transfer function. The LAPLACE part can be used to verify that a frequency response of a particular circuit corresponds to a specified transfer function.

CHAPTER 7

Fourier Series

The steady-state response of a linear circuit to a periodic input is another periodic signal. The period of the response is equal to the period of the input. Frequently, it is useful to represent periodic signals using the Fourier series. The Fourier series representation of a periodic signal $f(t)$ can be expressed as

$$f(t) = c_0 + \sum_{n=1}^{\infty} c_n \cos\left(n\omega_0 t + \theta_n\right)$$

where ω_0 is called the fundamental frequency and is related to the period, T, of $f(t)$ by

$$\omega_0 = \frac{2\pi}{T}$$

The Fourier series representation of a particular period function $f(t)$ is described by specifying the values of the coefficients $c_0, c_1, c_2, c_3,\ldots$ and $\theta_1, \theta_2, \theta_3,\ldots$ corresponding to that $f(t)$. For example, consider the periodic function $v_i(t)$ shown in Figure 7.1b. The

137

period of this function is 16 ms so the fundamental frequency is $\omega_0 = 125\pi$ rad/s. With some effort, the coefficients of the Fourier series of $v_i(t)$ can be determined to be

$$c_0 = -2, \quad c_n = \begin{cases} \dfrac{32}{n^2\pi^2} & \text{odd } n \\ 0 & \text{even } n \end{cases} \quad \text{and} \quad \theta_n = \begin{cases} \dfrac{-n\pi}{4} & \text{odd } n \\ 0 & \text{even } n \end{cases}$$

Therefore $v_i(t)$ is represented by the Fourier series

$$v_i(t) = -2 + \sum_{\substack{n=1 \\ \text{odd } n}}^{\infty} \frac{32}{n^2\pi^2} \cos\left(125\,n\pi t - \frac{n\pi}{4}\right) \quad V$$

$$\begin{aligned} = -2 \ &+ 3.242 \cos(393t - 45°) \\ &+ 0.36 \cos(1178t - 135°) \\ &+ 0.13 \cos(1963t - 225°) \\ &+ \ldots \end{aligned} \tag{7.1}$$

PSpice provides an easy way to calculate the values of the Fourier series coefficients c_0, c_1, c_2, c_3,... and θ_1, θ_2, θ_3,...corresponding to a periodic current or voltage.

Example 7.1 Consider the circuit shown in Figure 7.1a. The input to this circuit is the voltage of the voltage source, $v_i(t)$. The output of the circuit is the voltage, $v_o(t)$, across the 10 kΩ resistor. Both $v_i(t)$ and $v_o(t)$ are periodic voltages. The input, $v_i(t)$, is the periodic voltage shown in Figure 7.1b. The output, $v_o(t)$, will also be a periodic voltage. We will use PSpice to represent both $v_i(t)$ and $v_o(t)$ by Fourier series.

We will need to do three things:

1.   Represent the input voltage using one of the voltage sources shown in Table 4.1.

2.   Simulate the circuit in the time domain for a time that is long enough to include one full period after all transients have died out.

3.   Request that the Fourier series coefficients be calculated and printed in the PSpice output file.

Step 1 Formulate a circuit analysis problem.

Represent both $v_i(t)$ and $v_o(t)$, the input and output of the circuit shown in Figure 7.1, by their Fourier series.

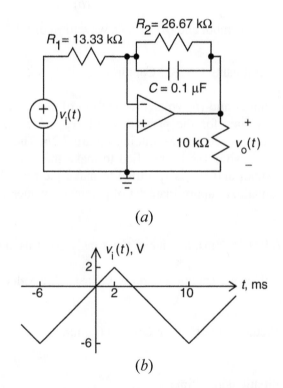

(a)

(b)

Figure 7.1 (*a*) A circuit and (*b*) a periodic input voltage.

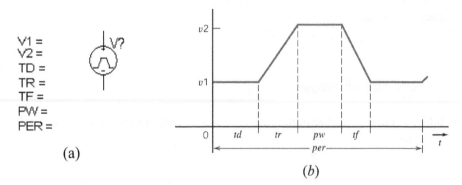

V1 =
V2 =
TD =
TR =
TF =
PW =
PER =

(a)

(b)

Figure 7.2 The (a) symbol and (b) voltage waveform of a VPULSE part.

Step 2 Describe the circuit using OrCAD Capture.

Figure 7.2 shows (a) symbol and (b) voltage waveform of a VPULSE part. This part is specified by providing values for the parameters $v1$, $v2$, td, tr, tf, pw and per. The meaning of each parameter is seen by examining Figure 7.2b. The pulse waveform will simulate the triangle wave when pw is specified to make the time that voltage remains equal to $v2$ negligibly small and per is specified to make the time that voltage remains equal to $v1$ negligibly small. An appropriate set of parameter values to simulate the input voltage, $v_i(t)$, is

$$v1 = 2 \text{ V}, v2 = -6 \text{ V}, td = 2 \text{ ms}, tr = 8 \text{ ms}, tf = 8 \text{ ms}, pw = 1 \text{ ns and } per = 16 \text{ ms.}$$

(PSpice requires $pw > 0$ so we cannot use $pw = 0$. Instead, a value much smaller than both tr and tf is used.)

Figure 7.3 shows the circuit as described in OrCAD Capture.

Step 3 Simulate the circuit using PSpice.

Select **PSpice / New Simulation Profile** from the OrCAD Capture menus to pop up the "New Simulation" dialog box. Specify a simulation name, then select "Create" to pop up the "Simulation Settings" dialog box as shown in Figure 7.4. Select "Time Domain(Transient)" as the analysis type. Specify the "Run to time" as 64 ms to run the simulation for four full periods of the input waveform.

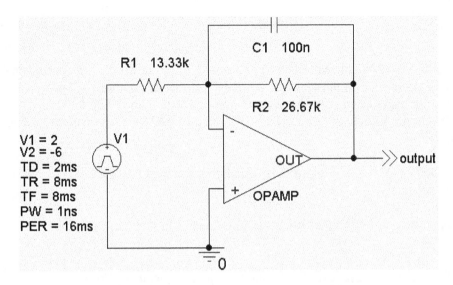

Figure 7.3 The circuit as described in OrCAD Capture.

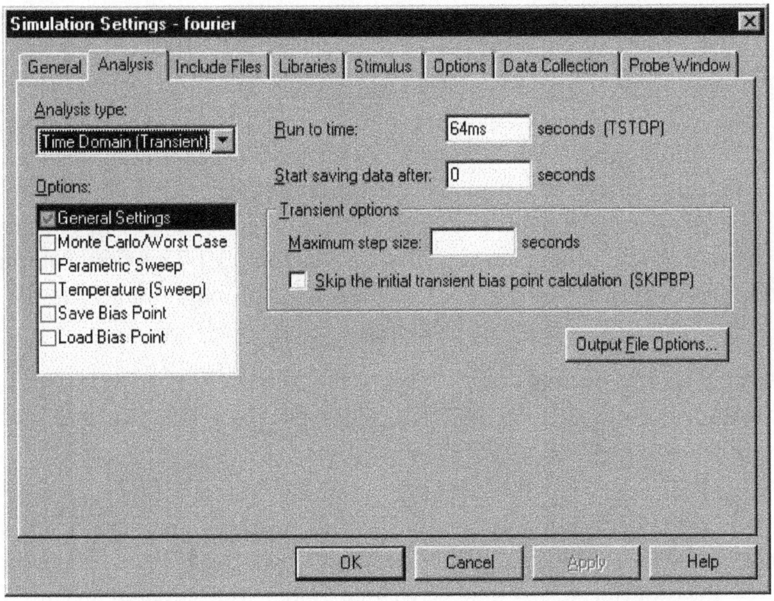

Figure 7.4 The "Simulation Settings" dialog box.

Left-click the Output File Options button to pop up the "Transient Output File Options" dialog box shown in Figure 7.5. Enter the fundamental frequency, ω_0, using units of Hertz, in the text box labeled "Center Frequency." Enter the number of coefficients c_1, c_2, c_3,... that are desired in the text box labeled "Number of Harmonics." Enter the PSpice names for voltages or currents that are to be represented by their Fourier series in text box labeled "Output Variables." Left-click "OK" to close the "Transient Output File Options" dialog box.

Finally, left-click "OK" to close the "Simulation Settings" dialog box and return to the Capture screen.

Select **PSpice / Run** from the OrCAD Capture menus to run the simulation.

Step 4 Display the results of the simulation, for example, using Probe.

After a successful Time Domain(Transient) simulation, Probe, the graphical post-processor for PSpice, will open automatically in a Schematics window. Select **View / Output File** from the Schematics menus. Scroll through the output file to find the Fourier coefficients of the input voltage shown in Figure 7.6. (PSpice changed the name of the input voltage. We used the name V(V1:+) in text box labeled "Output Variables" in the "Transient Output File Options" dialog box in Figure 7.5. Nonetheless, PSpice used the name V(N00230) in Figure 7.6.) The table in Figure 7.6 has 6 columns and eight rows. The eight rows correspond to the eight coefficients c_1, c_2, c_3,... c_8. (There are 8 rows because 8 was the number entered in the text box labeled "Number of Harmonics" in the "Transient Output File Options" dialog box in Figure 7.5.) The first column labels the rows with the subscripts, n, of these coefficients. The second column lists the frequencies $n\omega_0$, using units of Hertz. The third column lists the coefficients c_1, c_2, c_3,... c_8. The fourth column lists the normalized coefficients $c_1/c_1=1$, c_2/c_1, c_3/c_1,... c_8/c_1. The fifth column lists the phase angles θ_1, θ_2, θ_3,... θ_8. The sixth column lists the normalized coefficients $\theta_1-\theta_1=0$, $\theta_2-\theta_1$, $\theta_3-\theta_1$,... $\theta_8-\theta_1$.

We expect the even coefficients, c_2, c_4, c_6,... c_8 to be zero. They are much smaller than the odd coefficients so we will interpret them to be equal zero. The coefficient c_0 is the dc component of the Fourier series and is written above the table in Figure 7.5. Finally, PSpice represents the Fourier series using sine instead of cosine, so the coefficients in Figure 7.6 indicate that $v_i(t)$ is represented by the Fourier series

$$v_i(t) = -2.000199 + 3.242 \sin(393t + 45°)$$
$$+ 0.3602 \sin(1178t - 45°)$$
$$+ 0.1297 \sin(1963t - 135°)$$
$$+ 0.06613 \sin(2749t + 135°) + \ldots$$

We can represent the series using cosine by subtracting 90° from each phase angle. Then

$$v_i(t) = -2.000199 + 3.242 \cos(393t - 45°)$$
$$+ 0.3602 \cos(1178t - 135°)$$
$$+ 0.1297 \cos(1963t - 225°)$$
$$+ 0.06613 \cos(2749t + 45°) + \ldots$$

(7.2)

Scroll through the output file to find the Fourier coefficients of the output voltage shown in Figure 7.7. Figure 7.7 indicates that the Fourier series of $v_o(t)$ is

$$v_o(t) = 4.001551 + 4.444 \cos(393t + 88.4°)$$
$$+ 0.2112 \cos(1178t - 24.06°)$$
$$+ 0.04794 \cos(1963t - 118.8°)$$
$$+ 0.02040 \cos(2749t - 227°) + \ldots$$

(7.3)

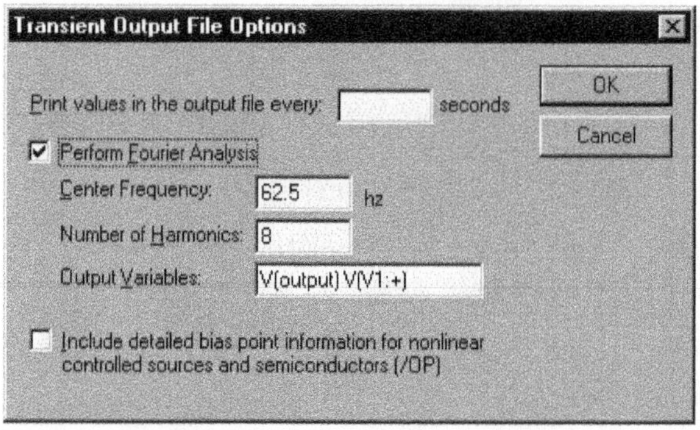

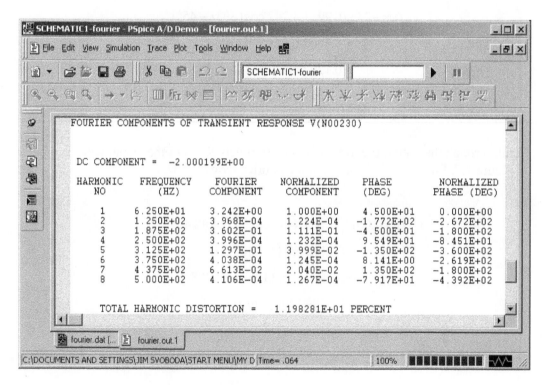

Figure 7.5 Requesting calculation of the Fourier series coefficients.

FOURIER COMPONENTS OF TRANSIENT RESPONSE V(N00230)

DC COMPONENT = -2.000199E+00

HARMONIC NO	FREQUENCY (HZ)	FOURIER COMPONENT	NORMALIZED COMPONENT	PHASE (DEG)	NORMALIZED PHASE (DEG)
1	6.250E+01	3.242E+00	1.000E+00	4.500E+01	0.000E+00
2	1.250E+02	3.968E-04	1.224E-04	-1.772E+02	-2.672E+02
3	1.875E+02	3.602E-01	1.111E-01	-4.500E+01	-1.800E+02
4	2.500E+02	3.996E-04	1.232E-04	9.549E+01	-8.451E+01
5	3.125E+02	1.297E-01	3.999E-02	-1.350E+02	-3.600E+02
6	3.750E+02	4.038E-04	1.245E-04	8.141E+00	-2.619E+02
7	4.375E+02	6.613E-02	2.040E-02	1.350E+02	-1.800E+02
8	5.000E+02	4.106E-04	1.267E-04	-7.917E+01	-4.392E+02

TOTAL HARMONIC DISTORTION = 1.198281E+01 PERCENT

Figure 7.6 The coefficients of the Fourier series of $v_i(t)$.

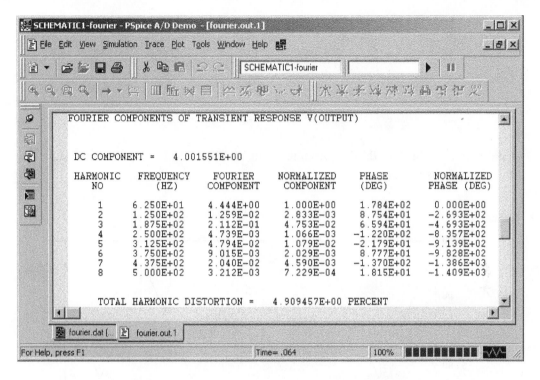

Figure 7.7 The coefficients of the Fourier series of $v_o(t)$.

Step 5 Verify that the simulation results are correct.

The Fourier series of $v_i(t)$ calculated using PSpice and given in Equation 7.2 agrees with the Fourier series of $v_i(t)$ given in Equation 7.1.

The network function of the circuit is

$$\mathbf{H}(\omega) = \frac{\mathbf{V}_o(\omega)}{\mathbf{V}_s(\omega)} = -\frac{R_2}{R_1 + j\omega C R_1 R_2} = \frac{-2}{1 + j\dfrac{\omega}{375}}$$

Consider the third harmonic, $n = 3$. The frequency is $\omega = n\omega_0 = 3\omega_0 = 1178$ rad/s. The phasor of the third harmonic term in the Fourier series of $v_o(t)$ is calculated as

$$\mathbf{V}_o\left(3\omega_0\right) = \frac{-2}{1+j\dfrac{3\omega_0}{375}} \times \mathbf{V}_s\left(3\omega_0\right) = \frac{-2}{1+j\dfrac{1178}{375}} \times \left(0.36\angle-135\right)$$

$$= 0.219\angle-27.3$$

which agrees with the third harmonic term, $0.211\angle-24.1°$ in Equation 7.3. The simulation results are correct.

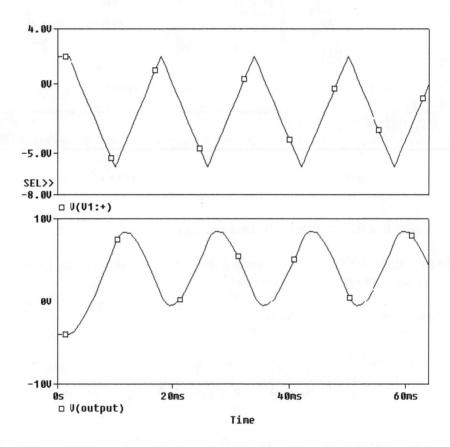

Figure 7.8 The input, $v_i(t)$, and response, $v_o(t)$, of the circuit in Figure 7.1.

Step 6 Report the answer to the circuit analysis problem.

The input and output voltages, $v_i(t)$ and $v_o(t)$, of the circuit shown in Figure 7.1*a* are represented by Fourier series given in Equations 7.2 and 7.3.

7.2 SUMMARY

PSpice provides an easy, three step procedure to calculate the values of the Fourier series coefficients of a periodic current or voltage.

1. Represent the input voltage using one of the voltage sources shown in Table 4.1.

2. Simulate the circuit in the time domain for a time that is long enough to have one full period after all transients have died out.

3. Request that the Fourier series coefficients be calculated and printed in the PSpice output file.

Appendix

Table A.1 PSpice parts and the libraries in which they are found

Symbol	Description	PSpice Name	Library
C? $-|\vdash-$ 1n	Capacitor	C	ANALOG
E?	VCVS (Voltage Controlled Voltage Source)	E	ANALOG
F?	CCCS (Current Controlled Current Source)	F	ANALOG
G?	VCCS (Voltage Controlled Current Source)	G	ANALOG

Symbol	Description	PSpice Name	Library
	Ground (reference node)	0	SOURCE
H? ... H	CCVS (Current Controlled Voltage Source)	H	ANALOG
1Aac 0Adc	AC Current Source	IAC	SOURCE
0Adc	DC Current Source	IDC	SOURCE
I1 = I2 = TD1 = TC1 = TD2 = TC2 =	Exponential Current Source	IEXP	SOURCE

Symbol	Description	PSpice Name	Library
	Piecewise Linear Current Source	IPWL	SOURCE
IOFF = IAMPL = FREQ =	Sinusoidal Current Source	ISIN	SOURCE
K? K_Linear COUPLING = 1	Coupling Coefficient	K_Linear	ANALOG
L? 1 ⌒⌒⌒ 2 10uH	Inductor	L	ANALOG
$\dfrac{1}{1+s}$	Transfer Function	LAPLACE	ABM
U? OP AMP	Op Amp	OPAMP	ANALOG

Symbol	Description	PSpice Name	Library
OFFPAGELEFT-R	Off page connector	OFFPAGELEFT-R	CAPSYM
PARAMETERS:	Global Parameter	PARAM	SPECIAL
IPRINT	Current printer (Ammeter)	IPRINT	SPECIAL
	Node voltage printer (voltmeter)	VPRINT1	SPECIAL
	Element voltage printer (voltmeter)	VPRINT2	SPECIAL
R? 1k	resistor	R	ANALOG

Symbol	Description	PSpice Name	Library
TCLOSE = 0 1 ⟋ 2 U?	open switch, will close at t = TCLOSE	Sw_tClose	EVAL
TOPEN = 0 1 ⟋ 2 U?	closed switch, will open at t = TOPEN	Sw_tOpen	EVAL
S? S VOFF = 0.0V VON = 1.0V	voltage controlled switch	S	ANALOG
TX?	transformer	XFMR_ LINEAR	ANALOG

Symbol	Description	PSpice Name	Library
1Vac ⊕ 0Vdc V?	AC voltage	VAC	SOURCE
0Vdc ⊕ V?	DC voltage source	VDC	SOURCE
V1 = V2 = TD1 = TC1 = TD2 = TC2 = V?	Exponential Voltage Source	VEXP	SOURCE
V?	Piecewise Linear Voltage Source	VPWL	SOURCE
VOFF = VAMPL = FREQ = V?	Sinusoidal Voltage Source	VSIN	SOURCE

Index